ROBOTS IN BRITISH INDUSTRY

Expectations and Experience

JIM NORTHCOTT with
Colin Brown, Ian Christie,
Michael Sweeney
and Annette Walling

PSI Research Report No. 660

Policy Studies Institute
100 Park Village East
London NW1 3SR

D
629.892
ROB.

© Policy Studies Institute 1986

All rights reserved. No part of this publication may be reproduced, stored in a retrieval system, or transmitted, in any form or by any means, electronic, electrical, chemical, mechanical, optical, photocopying, recording or otherwise, without the prior permission of the copyright owner.

PSI publications are obtainable from all good bookshops, or by visiting the Institute at 100, Park Village East, London NW1 3SR (01-387 2171).

Sales Representation: Frances Pinter (Publishers) Ltd. Orders to Marston Book Services, P.O. Box 87, Oxford, OX4 1LB.

ISBN 0-85374-323-1

Published by Policy Studies Institute, 100 Park Village East, London NW1 3SR.

Printed by Blackmore Press, Longmead, Shaftesbury, Dorset

ROBOTS IN BRITISH INDUSTRY

CONTENTS

	PAGE
SUMMARY	1
Extent of use	2
The project	2
Characteristics of robot users	3
Introduction of robots	5
Government support for using robots	6
Number of robots	6
Applications	7
Training, employment and industrial relations	8
Difficulties	10
Benefits	10
Factors in success	11
Plans and prospects	14
Policy implications	15

1 INTRODUCTION
    The study   19
    The report   20
      20

2 EXTENT OF USE
    Defining robots   23
    Total number of robot users and robots   23
    Growth in number of robot users and robots   24
    Extent of diffusion   24
      25

3 CHARACTERISTICS OF ROBOT USERS
    Size   27
    Industry   27
    Shiftwork   27
      28

4 INTRODUCTION OF ROBOTS
    Source of initiative   29
    Feasibility studies   29
    Approach to installation   30
    Time needed for installation   31
      32

5 GOVERNMENT SUPPORT FOR USING ROBOTS
    Grants for feasibility studies   33
    Grants for purchase of robots and development of applications   33
      34

Contents

6 NUMBER OF ROBOTS 37
Number of robots per plant 37
Type of company 37
Size 38
Industry 38
Hours of use and shifts 38

7 APPLICATIONS 39
Existing robot users 39
Prospective users 40

8 TRAINING, EMPLOYMENT, INDUSTRIAL RELATIONS 41
Attitudes of workers affected 41
Effects on employment 42
Training 43

9 DIFFICULTIES 45
Difficulties expected in advance 45
Difficulties actually experienced 46
Costs 46
Installation, expertise and after-sales support 47
Reliability and maintenance 48

10 BENEFITS 49
Main benefits experienced 49
Effects of plant size 49
Effects of industry 49
Effects of length of use 50
Effects of number of shifts 50
Effects of number of robots 50
Profitability 51
Intention to buy more robots 51
Overall value 51

11 FACTORS IN SUCCESS 53
Plant characteristics 53
Introduction of robots 54
Number and application of robots 56
Benefits and difficulties 56

12 PLANS AND PROSPECTS 59
Existing robot users 59
Prospective robot users 60
Growth prospects 60
Requirements of robot users 62

13 POLICY IMPLICATIONS 63
Robot users 63
Robot suppliers 66
Government 67

APPENDIX I PARTICULARS OF THE STUDY 71
Form and purpose of the study 71
Timing 71
Questionnaires 71
Robot users sample 72
Non-users sample 72
Despatch and response 73
Total numbers of robot users and robots in UK 73
Future numbers of robot users 74
Case-study interviews 75
Presentation 75

APPENDIX II POSTAL SURVEY QUESTIONNAIRES 77

APPENDIX III DEPARTMENT OF TRADE AND INDUSTRY SUPPORT FOR ROBOTS 89

APPENDIX IV GLOSSARY OF TECHNICAL TERMS 91

APPENDIX V PSI'S MICROELECTRONICS APPLICATIONS RESEARCH PROGRAMME 93

TABLES 98

TABLES

Note: The 'Q' numbers refer to the numbers of the questions in the main questionnaire to which the tables relate. Thus Q.4 × Q.6 indicates Question 4 tabulated against Question 6. 'NQ' numbers refer to the numbers of the questions in the separate questionnaire sent to prospective users.

EXTENT OF USE (pages 98-100)

1. Extent of use of robots compared with use of other kinds of automated manufacturing technology
2. Robot users' use of other advanced manufacturing technology
3. Extent of use of robots by employment size of user
4. Extent of use of robots by industry
5. Extent of use of robots by type of company
6. Extent of use of robots by region

CHARACTERISTICS OF ROBOT USERS (pages 101-105)

7. Distribution of robot users by employment size and stage of use of first robot (Q.3 × Q.13)
8. Distribution of robot users by company turnover and stage of use of first robot (Q.5 × Q.13)
9. Distribution of robot users by number of plants and stage of use of first robot (Q.2 × Q.13)
10. Distribution of robot users by industry and stage of use of first robot (Q.6 × Q.13)
11. Distribution of robot users by industry group and by employment size, turnover and number of plants (Q.1, 2, 3, 5 × Q.6)
12. Distribution of robot users by number of shifts worked and stage of use of first robot (Q.4 × Q.13)
13. Distribution of robot users between those which are and are not robot suppliers by stage of use of first robot (Q.7 × Q.13)

INTRODUCTION OF ROBOTS (pages 105-119)

14. Stage of use of first robot by type of company (Q.13 × Q.1)
15. Stage of use of first robot by year first robot acquired (Q.13 × Q.14)
16. Year first robot acquired by stage of use of first robot (Q.14 × Q.13)
17. Year first robot acquired by employment size of plant (Q.14 × Q.3)
18. Year first robot acquired by number of plants of company (Q.14 × Q.2)
19. Year first robot acquired by number of shifts a day (Q.14 × Q.4)
20. Year first robot acquired by industry group (Q.14 × Q.6)
21. Origin of idea and place of decision to use robots by stage of use of first robot (Q.9 × Q.13)
22. Origin of idea and place of decision to use robots by employment size of plant (Q.9 × Q.3)
23. Origin of idea and place of decision to use robots by number of plants of company (Q.9 × Q.2)
24. Origin of idea and place of decision to use robots by industry group (Q.9 × Q.6)
25. Origin of idea and place of decision to use robots by number of robots in use (Q.9 × Q.17)
26. Organisation undertaking feasibility study by stage of use of first robot (Q.10 × Q.13)
27. Organisation undertaking feasibility study by employment size of plant (Q.10 × Q.3)
28. Organisation undertaking feasibility study by number of robots in use (Q.10 × Q.17)
29. Rating of consultant's report by stage of use of first robot (Q.11 × Q.13)
30. Approach to installation of first robot by stage of use of first robot (Q.16 × Q.13)
31. Approach to installation of first robot by type of feasibility study (Q.16 × Q.10)
32. Approach to installation of first robot by industry group (Q.16 × Q.6)
33. Number of months needed to get first robot into commercial production by stage of use of first robot (Q.15 × Q.13)
34. Number of months needed to get first robot into commercial production by type of feasibility study (Q.15 × Q.8)
35. Number of months needed to get first robot into commercial production by industry group (Q.15 × Q.6)

GOVERNMENT SUPPORT FOR USING ROBOTS (pages 120-133)

36. Awareness of government grants for feasibility studies by whether received support for development of robot application (Q.12)
37. Awareness of government grants for feasibility studies by type of feasibility study (Q.12 × Q.10)
38. Awareness of government grants for feasibility studies by type of company (Q.12 × Q.1)
39. Awareness of government grants for feasibility studies by employment size of plant (Q.12 × Q.3)
40. Awareness of government grants for feasibility studies by industry group (Q.12 × Q.1)
41. Awareness of government grants for feasibility studies by number of robots in use (Q.12 × Q.17)
42. Time of awareness of support to help use robots of plants receiving grants by employment size of plant (Q.36 × Q.3)
43. Time of awareness of support to help use robots of plants receiving grants by industry group (Q.36 × Q.6)
44. Source of awareness of support to help use robots of plants receiving grants by employment size of plant (Q.37 × Q.3)
45. Source of awareness of support to help robots of plant receiving grants by industry group (Q.37 × Q.6)
46. Reasons of non-applicants for not applying for government support to help with the purchase and development costs of robots by employment size of plant (Q.36 × Q.3)
47. Reasons of non-applicants for not applying for government support to help with the purchase and development costs of robots by industry group (Q.36 × Q.6)
48. Action by grant recipients in the absence of support for use of robots by employment size of plant (Q.38 × Q.3)
49. Action by grant recipients in the absence of support for use of robots by industry group (Q.38 × Q.6)
50. Year first robot acquired by whether received support for development of robot application (Q.14)
51. Willingness of grant recipients to undertake a further similar project without a grant by employment size of plant (Q.39 × Q.3)
52. Willingness of grant recipients to undertake a further similar project without a grant by industry group (Q.39 × Q.6)
53. Stage of use of first robot by whether received support for development of robot application (Q.13)
54. Number of robots per plant by whether received support for development of robot application (Q.17)
55. Non-users' awareness and use of grants for feasibility studies by plans to use robots (NQ.10 × Q.8)
56. Non-users' plans to use robots in next two years by awareness of grants for robots (NQ.8 × NQ.10)
57. Non-users' awareness of grants for investment in robots by employment size of plant (NQ.11 × Q.3)

NUMBER OF ROBOTS (pages 134-139)

58. Number of robots per plant by number of robots in use (Q.17)
59. Number of robots per plant and total number of robots by year first robot acquired (Q.17 × Q.14)
60. Number of robots per plant and total number of robots by type of company (Q.17 × Q.1)
61. Number of robots per plant and total number of robots by number of plants of company (Q.17 × Q.2)
62. Number of robots per plant and total number of robots by employment size of plant (Q.17 × Q.3)
63. Number of robots per plant by industry group (Q.17 × Q.1)
64. Number of robots per plant and total number of robots by number of shifts a day (Q.17 × Q.4)
65. Hours a week robot scheduled to be running by number of shifts a day (Q.20 × Q.4)
66. Hours a week robot scheduled to be running and number of shifts worked by employment size of plant (Q.20, 4 × Q.3)
67. Hours a week robot scheduled to be running and number of shifts worked by industry group (Q.20, 4 × Q.6)

APPLICATIONS
(pages 140-144)

68. Robot applications (Q.19)
69. Robot applications by employment size of plant (Q.19 × Q.3)
70. Robot applications' distribution between plants of different employment sizes (Q.3 × Q.19)
71. Robot applications by industry group (Q.19 × Q.6)
72. Robot applications' distribution between industry groups (Q.6 × Q.19)
73. Robot applications by total number of robots in use in plant (Q.19 × Q.17)
74. Robot applications' distribution between plants with different numbers of robots (Q.17 × Q.19)
75. Robot applications envisaged by non-users (NQ.12 × NQ.8)

TRAINING, EMPLOYMENT, INDUSTRIAL RELATIONS
(pages 145-157)

76. Attitude of the workers directly affected by the introduction of robots by stage of use of first robot (Q.27 × Q.13)
77. Attitude of the workers directly affected by the introduction of robots by number of robots in use (Q.27 × Q.17)
78. Attitude of the workers directly affected by the introduction of robots by whether there was advance consultation and whether there were negotiations with the unions (Q.27 × Q.28, 29)
79. Attitude of the workers directly affected by the introduction of robots by changes in employment as a direct result of the introduction of robots (Q.27 × Q.30)
80. Consultation with those directly affected and negotiation with trade unions by stage of use of first robot (Q.28, 29 × Q.13)
81. Consultation with those directly affected and negotiation with trade unions by changes in employment as a direct result of the introduction of robots (Q.28, 29 × Q.30)
82. Changes in employment at the plant as a direct result of the introduction of robots by stage of use of first robot (Q.30 × Q.13)
83. Changes in employment as a direct result of the introduction of robots by type of company (Q.30 × Q.1)
84. Changes in employment as a direct result of the introduction of robots by employment size of plant (Q.30 × Q.3)
85. Changes in employment as a direct result of the introduction of robots by industry group (Q.30 × Q.6)
86. Changes in employment as a direct result of the introduction of robots by year first robot acquired (Q.30 × Q.14)
87. Changes in employment as a direct result of the introduction of robots by number of robots in use (Q.30 × Q.17)
88. Numbers already been on robot training courses and numbers planned to go in next two years by type of organisation providing training course (Q.31)
89. Numbers already been on robot training courses and numbers planned to go in next two years by employment size of plant (Q.31 × Q.3)
90. Numbers already been on robot training courses and numbers planned to go in next two years by industry group (Q.31 × Q.6)
91. Numbers already been on training courses and numbers planned to go in next two years by number of robots in use (Q.31 × Q.17)
92. Extent of satisfaction with training arrangements by employment size of plant (Q.32 × Q.3)
93. Extent of satisfaction with training arrangements by industry group (Q.32 × Q.6)

DIFFICULTIES
(pages 158-177)

94. Difficulties and disadvantages expected by robot users before going into production compared with those actually experienced after (Q.24)
95. Main difficulties and disadvantages expected by robot users compared with those actually experienced by stage of use of first robot (Q.24 × Q.13)
96. Main difficulties and disadvantages expected by robot users compared with those actually experienced by employment size of plant (Q.24 × Q.3)

97. Main difficulties and disadvantages expected by robot users compared with those actually experienced by industry group (Q.24 × Q.6)
98. Main difficulties and disadvantages expected by robot users compared with those actually experienced by type of feasibility study (Q.24 × Q.10)
99. Main difficulties and disadvantages expected by robot users compared with those actually experienced by year first robot acquired (Q.24 × Q.14)
100. Difficulties and disadvantages experienced by robot users by stage of use of first robot (Q.24 × Q.13)
101. Difficulties and disadvantages experienced by robot users by employment size of plant (Q.24 × Q.3)
102. Difficulties and disadvantages experienced by robot users by industry group (Q.24 × Q.6)
103. Difficulties and disadvantages experienced by robot users by type of feasibility study (Q.24 × Q.10)
104. Difficulties and disadvantages experienced by robot users by year first robot acquired (Q.24 × Q.14)
105. Main difficulties and disadvantages experienced by robot users by number of robots in use (Q.24 × Q.17)
106. Difficulties and disadvantages expected by non-users compared with difficulties and disadvantages expected and actually experienced by existing robot users (NQ.14 × NQ.8)
107. Disadvantages and problems experienced by robot users compared with users of other microelectronics-based production technologies
108. Price of robots by stage of use (Q.18 × Q.13)
109. Price of robots by industry group (Q.18 × Q.6)
110. Price of robots by type of feasibility study (Q.18 × Q.10)
111. Downtime of robots compared with expectations by stage of use of first robot (Q.21 × Q.13)
112. Downtime of robots compared with expectations by industry group (Q.21 × Q.6)
113. Downtime of robots compared with expectations by type of feasibility study (Q.21 × Q.10)
114. Downtime of robots compared with expectations by year first robot acquired (Q.21 × Q.14)
115. Downtime of robots compared with expectations by number of robots in use (Q.21 × Q.17)
116. Frequency of robot downtime by type of cause (Q.22)
117. Frequency of robot downtime by type of cause and employment size of plant (Q.22 × Q.3)
118. Frequency of robot downtime by type of cause and industry group (Q.22 × Q.6)
119. Frequency of robot downtime by type of cause and type of feasibility study (Q.22 × Q.10)
120. Frequency of downtime by type of cause and number of robots in use (Q.22 × Q.17)

BENEFITS (pages 178-186)

121. Benefits expected by robot users before going into production compared with those actually experienced after (Q.23)
122. Benefits expected by non-users compared with benefits expected and actually experienced by existing robot users (NQ.15 × NQ.8)
123. Benefits experienced by robot users by employment size of plant (Q.23 × Q.3)
124. Benefits experienced by robot users by industry group (Q.23 × Q.6)
125. Benefits experienced by robot users by stage of use of first robot (Q.23 × Q.13)
126. Benefits experienced by robot users by year first robot acquired (Q.23 × Q.14)
127. Benefits experienced by robot users by number of shifts a day (Q.23 × Q.4)
128. Benefits experienced by robot users by number of robots in use (Q.23 × Q.17)
129. Worthwhileness and profitability of use of robots by number of robots in use (Q.25, 26, 27)

FACTORS IN SUCCESS (pages 187-190)

130. Factors in success: plant characteristics
131. Factors in success: approach to introduction of robots
132. Factors in success: number and application of robots
133. Factors in success: difficulties and benefits experienced

PLANS AND PROSPECTS
(pages 191-210)

134. Plans to acquire more robots in next two years by industry group (Q.33 × Q.6)
135. Plans to acquire more robots in next two years by type of company (Q.33 × Q.1)
136. Plans to acquire more robots in next two years by employment size of plant (Q.33 × Q.3)
137. Plans to acquire more robots in next two years by number of shifts a day (Q.33 × Q.4)
138. Plans to acquire more robots in next two years by year first robot acquired (Q.33 × Q.14)
139. Plans to acquire more robots by number of robots in use now (Q.33 × Q.17)
140. Number of robots per plant now by number of additional robots planned in next two years (Q.17 × Q.33)
141. Plans to acquire more robots in next two years by whether robot downtime more or less than expected (Q.33 × Q.21)
142. Plans to acquire more robots in next two years by extent use of robots worthwhile and profitable (Q.33 × Q.25, 26)
143. Non-users' robot feasibility studies by employment size of plant (NQ.9 × Q.3)
144. Non-users' robot feasibility studies by number of shifts a day (NQ.9 × Q.4)
145. Non-users' robot feasibility studies by industry group (NQ.9 × Q.6)
146. Non-users' plans to use robots in next two years by whether feasibility study undertaken (NQ.8 × Q.9)
147. Non-users' plans to use robots in next two years by employment size of plant (NQ.8 × Q.3)
148. Non-users' plans to use robots in next two years by number of plants of company (NQ.8 × Q.2)
149. Non-users' plans to use robots in next two years by number of shifts a day (NQ.8 × Q.4)
150. Non-users' plans to use robots in the future by industry group (NQ.8 × Q.6)
151. Growth in total number of robots and robot users in UK on alternative assumptions by industry group
152. Growth in total number of robots and robot users in UK on alternative assumptions by employment size of plant
153. Plans to acquire more robots in next two years by degree of sophistication of new robot (Q.33 × Q.34)
154. Improvements robot users consider would help effective use in the future by employment size of plant (Q.35 × Q.3)
155. Improvements robot users consider would help effective use in the future by industry group (Q.35 × Q.6)

PARTICULARS OF SURVEY SAMPLES
(pages 211-215)

A. Basis of postal survey: sources of sample and rates of response
B. Sample of robot users by plant size and industry
C. Sample of non-users by plant size and industry
D. Comparison of the plant size distribution of the robot users in the survey sample with the UK distribution of robot users in the PSI survey of microelectronics in industry
E. Comparison of the distribution by industry of the robot users in the survey sample with the UK distribution of robot users in the PSI survey of microelectronics in industry

ACKNOWLEDGEMENTS

This report presents the work not of an individual but of a team. At PSI Annette Walling organised the postal surveys; Colin Brown undertook the computer analysis; Ian Christie and Michael Sweeney carried out and interpreted the case study interviews; and Jim Northcott designed and directed the study, prepared the tables and wrote most of the report. In addition we have had much help from others at PSI, including Nancy Walling who prepared the survey data for computing, and Rosemary Lewin who prepared the text for the printer.

We have also had important contributions from outside PSI. In particular, we are most grateful to James Fleck of the Department of Business Studies at Edinburgh University, who provided comprehensive background information and advice which was most useful in helping get the project off to a quick start, and to Prof. W. B. Heginbotham, OBE, DSC, F.Eng, whose comments on the report draft have been of great value.

We are also grateful to the organisations which contributed to the assembly of the large and representative samples for the surveys: to the Department of Trade and Industry for providing a list of the firms which received grants under the robot support scheme, which accounted for two-fifths of our total sample of robot users; to Cahners Exhibitions Limited for making available a relevant selection of the list of visitors to the 1985 Automan exhibition, which accounted for the overwhelming majority of our sample of non-users and also about one quarter of the sample of existing users; to the robot suppliers who entrusted to us lists of their customers, thereby adding usefully to the total number of users in the sample; and to the the British Robot Association whose company members provided a further addition to both samples.

We also appreciate the trouble taken by the 787 robot users and non-users in completing the postal questionnaires which provided the basic data for the surveys, and to the others who gave further valuable time to talk with us in the supplementary interviews.

We wish to thank Ingersoll Engineers for permission to draw on definitions in the glossary of their report *Integrated Manufacture* (IFS, 1985). This in turn draws on *How To Speak Automation*, published by General Electric in conjunction with Grant Publications in the USA.

Finally, we are greatly indebted to the Gatsby Charitable Foundation, the Department of Trade and Industry and the National Economic Development Office for providing the funding to enable the study to be undertaken, and to the British Robot Association for taking the original initiative which resulted in the project going ahead. We much appreciate the encouragement, advice and support received from all of them throughout the study, but wish to make it clear that none of them should be regarded as in any way responsible for the manner in which the study was conducted or for the information or opinions contained in the report.

The survey was undertaken as part of PSI's programme of research into applications of new technology. The programme was initiated by Sir Charles Carter and, under his chairmanship, has enjoyed the support of an Advisory Committee consisting of Dr. Gordon Fryers, Maurice Goldsmith, Sir Ieuan Maddock CBE, OBE, FRS, John Maddox, John Major, Dr. A. J. Pope and Sir Bruce Williams.

SUMMARY

Summary

INTRODUCTION

Robots are a vital element in factory automation. How much are they actually being used in British industry? In what ways and with what effects — for example on profitability and on jobs? What are the obstacles, and the benefits? What are the key factors in success?

To get the facts PSI has undertaken major surveys of robot users and prospective users from all sectors of manufacturing industry in the UK.

A simplified summary of the surveys' findings is given below. A fuller explanation of each topic is given in the main text, and detailed figures are provided in the main tables which follow it.

THE PROJECT

The study was undertaken by PSI on the initiative of the British Robot Association. Funding was provided by the Gatsby Charitable Foundation, the Department of Trade and Industry and the National Economic Development Office.

The study took the form of postal surveys of robot users and potential users, followed up by a series of case-study interviews with a selection of users, non-users, suppliers and others. The postal survey questionnaires were sent out at the end of 1985 and analysed by computer in the winter and spring of 1986, and the interviews were held in the spring of 1986.

Response rates were high for surveys of this kind. The questionnaire for users elicited a 57% response (representing about one third of all UK robot users); that for potential users gave a 41% response. This gave effective samples of 248 current users (about one third of the UK total) and 363 potential users. Mini-questionnaires were sent to firms which had not replied and these took the overall users' response to 73% and the potential users' response to 53%, bringing the total samples to 326 and 461 respectively for the key questions covered in the mini-questionnaire.

Full details of the survey samples, questionnaires, research methods and statistics are given in Appendix I and Tables A – E following it.

EXTENT OF USE
(Tables 1-6, 15-20)

The number of plants using robots in Britain increased by about 50 per cent a year between 1981 and 1984, but the rate of increase was smaller in 1985, and the total number of users in the UK at the beginning of 1986 was still only about 740 — less than one factory in every forty. The percentage of users is much higher in the large plants than in the small ones, and three times as high in the plants owned by overseas companies as in the British ones.

The British Robot Association estimates that the total number of robots in Britain at the end of 1985 was about 3,200 — less than the *increase* in the number in Germany in 1985 alone.

Summary

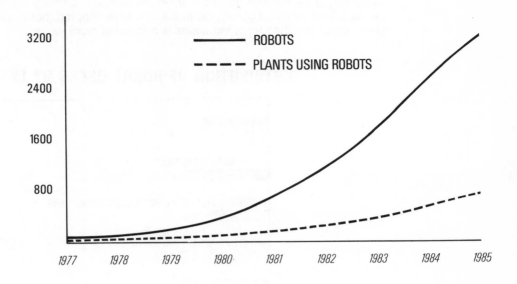

GROWTH IN USE OF ROBOTS

CHARACTERISTICS OF ROBOT USERS
Size
(Tables 7-9)

Robot users tend to be the larger plants: three-quarters of them are in plants employing over 200, one quarter have a turnover of more than £100 million, and one third are in companies with five or more plants.

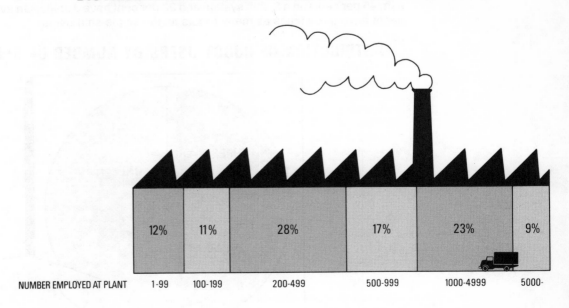

DISTRIBUTION OF ROBOT USERS BY EMPLOYMENT SIZE

NUMBER EMPLOYED AT PLANT	1-99	100-199	200-499	500-999	1000-4999	5000-
	12%	11%	28%	17%	23%	9%

Summary

Industry

(Tables 10, 11, 13)

About 60 per cent of all robot users are in the vehicle and engineering sectors, with a further 13 per cent in other metal goods. Other sectors have adopted robots more recently and there are some signs that the mix of industries is becoming more varied.

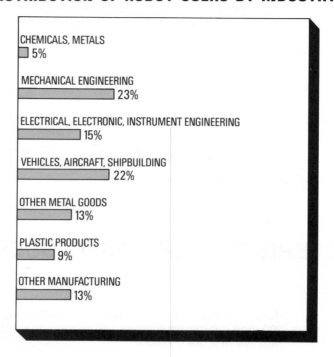

DISTRIBUTION OF ROBOT USERS BY INDUSTRY

Shiftwork

(Table 12)

Only 29 per cent of the plants with robots installed for commercial production work a single shift. 45 per cent run a 2-shift system and 22 per cent have 3 shifts. On average the 2- or 3- shift plants have three times as many robots as the single-shift users.

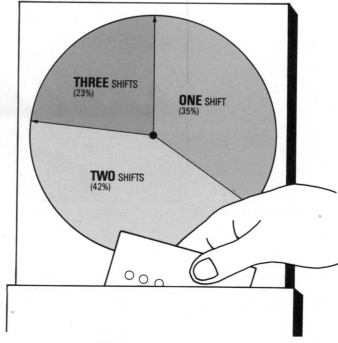

DISTRIBUTION OF ROBOT USERS BY NUMBER OF SHIFTS WORKED

Summary

INTRODUCTION OF ROBOTS

Source of initiative
(Tables 21-25)

For more than 40 per cent of all robot users the original idea came from plant management. A third of the users (mainly the smaller ones) report that the initiative came from the company board or head office.

In all kinds of organisation the final decision tends to be taken at a higher level than that at which the original idea was generated.

Feasibility studies
(Tables 26-29, 31)

The great majority of the plants using robots undertook a feasibility study before the introduction of robots. One in ten did without, but took half as long again as the others to get their first robot into production.

Of the potential users surveyed, 21 per cent have completed a feasibility study, but over half of the reports have recommended against using robots.

Approach to installation
(Tables 30-32)

Nearly half of the users sought to avoid installation problems by opting for a 'turnkey' installation — a complete package provided by the supplier — instead of buying just the robot and carrying out installation and development themselves.

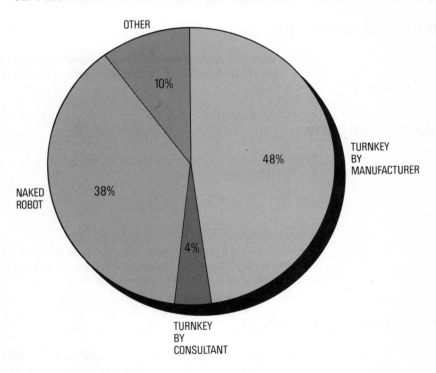

APPROACH TO INSTALLATION OF FIRST ROBOT

Time needed for installation
(Tables 33-35)

Nearly half of the first robots went into production work within three months and nearly three-quarters of them within six months. Installation time is usually much reduced for second and subsequent robots, especially when they are used for the same tasks as the first one. It has been well below average in the overseas-owned plants.

Summary

GOVERNMENT SUPPORT FOR USING ROBOTS

Grants for feasibility studies (Tables 36-41, 55)

Government grants are available towards the cost of feasibility studies. Although nearly all users carried out a study, only one in four received a grant towards the cost. One in seven of robot users and prospective users were unaware of the availability of grants, and over half of those who were aware did not apply. Only 7 per cent of applicants were turned down — these were mainly very large firms or plants using more than one robot already.

Grants for purchase and development (Tables 42-54)

Until June 1986 grants were also available towards the purchase costs of robots, and associated equipment and for development work. Just under half the robot users have received these grants.

Of the users who did not receive assistance, 26 per cent did not apply because they were not eligible, 14 per cent were unaware of the scheme and 22 per cent objected to the speed of processing or the conditions.

Recipients of the grants regard them as crucial or very important to their robot projects. Many have criticisms of the operation of the scheme but very few question the policy of support.

NUMBER OF ROBOTS

Number of robots per plant (Tables 58, 59)

43 per cent of the plants in the sample have only one robot; 18 per cent have two, and 16 per cent have more than five. The average number of robots per plant among all users in the UK is about 4.3.

Type of company (Table 60)

The overseas-owned companies in Britain have on average twice as many robots per plant as the British ones.

Size (Tables 61, 62)

Two-thirds of the plants employing less than 200 people have only one robot, and on average thay have only about two robots each. By contrast the plants employing over 1,000 people on average have about seven robots each.

Distribution by industry (Table 63)

Plants in the vehicles and plastics sector tend to use robots in larger numbers, with an average of about seven each. The vehicles sector accounts for over two-fifths of the UK's total robot population.

Hours of use and shifts (Tables 64-67)

The larger installations tend to be in the plants with more than one shift where the robots work much longer hours. The 8 per cent of users with more than ten robots each account between them for about 46 per cent of all the robots used and for about 64 per cent of all the hours worked per week by the robots.

Summary

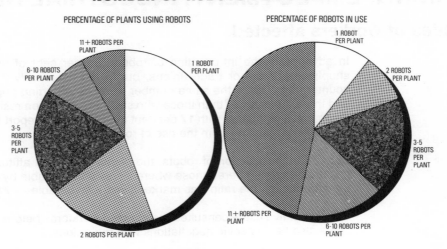

APPLICATIONS

Existing robot users
(Tables 68-74)

About one third of all the plants using robots use them for arc welding; about one sixth for assembly, machine loading, painting and coating, and for handling. In spot welding and injection moulding there are fewer users but they tend to use robots in much greater numbers with the result that more than 500 are used for each — more than in any other application.

Prospective robot users
(Table 75)

The applications most often envisaged by prospective users are assembly, machine loading and arc welding.

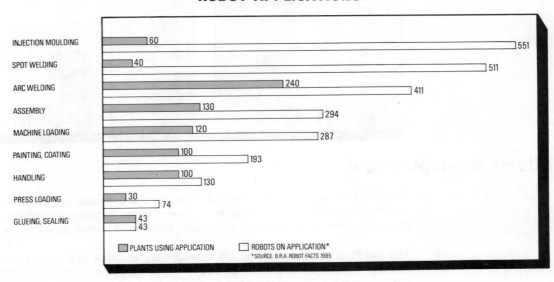

Summary

TRAINING, EMPLOYMENT, INDUSTRIAL RELATIONS

Attitudes of workers affected
(Tables 76-81, 94, 121)

In advance of the introduction of robots, 31 per cent of users expected opposition from shopfloor workers or trade unions, but in the event only 2 per cent of them actually encountered it. This is the same number as those reporting opposition from top management, and three times fewer than those who say there has been resistance from other groups in the company. It compares with 17 per cent of plants which report *better* labour relations as one of the *benefits* resulting from the use of robots.

Before the introduction of robots, the plants where the attitude of the workers affected was favourable outnumbered those where it was unfavourable by 42 per cent to 9 per cent. After the introduction the ratio was markedly *more* favourable — 71 per cent to 4 per cent.

Four out of five users consulted the affected workforce before introduction. About a quarter of users also held specific negotiations with the unions.

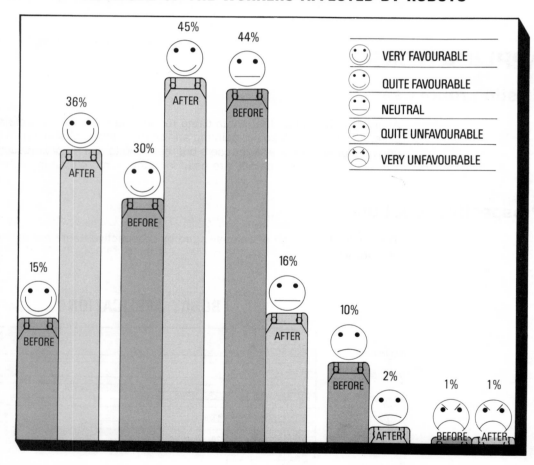

Effects on employment
(Tables 79, 81-87)

So far the introduction of robots has not had the dramatic impact on jobs which some have feared. Three-fifths of users report no change in the numbers employed as a result of their use of robots. One in four report a decrease, and one in twelve an increase due to robot use.

The net total decrease in the UK due directly to robot use is only about 700, shed mostly through natural wastage, or redeployment of displaced workers to other jobs in the same plant. However, job losses have been greater in the plants which have more robots and in those which have been using them for many years, and total losses are likely to become greater when more robots are used over a longer period.

Summary

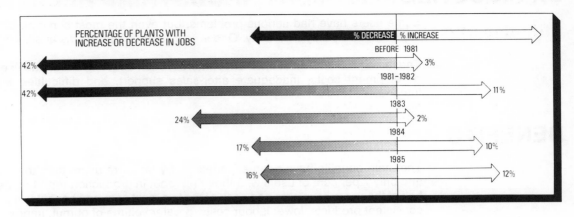

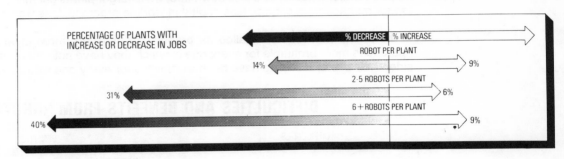

(Tables 88-93)

Training

Three-quarter of users have sent staff on training courses provided by the robot suppliers, and about a quarter provide in-house training. Most users appear to be satisfied with their training arrangements and there seem to be few with serious problems so far.

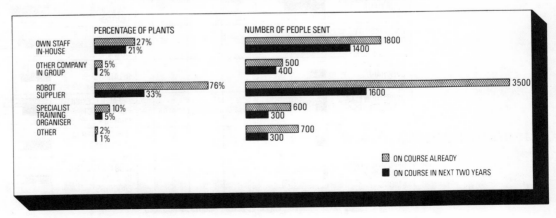

Summary

DIFFICULTIES
(Tables 94-105)

Some users have had serious problems, but even the most common difficulties have been experienced by only a minority. One in four report no problems at all.

The three problems most often encountered, each affecting about one user in three, are: high development costs, inadequate after-sales support, and difficulties with installation and integration.

BENEFITS
(121-129, 135)

The main benefits have been experienced by far more users than the main disadvantages, and only 7 per cent of users with their first robot in production report no benefits at all. About half of the users have obtained each of the five most common benefits: improved quality, more consistent products, lower labour costs, greater volume of output, improved work conditions, safety, increased technical expertise.

In general, smaller plants tend to get more technical benefits (improved quality, increased output, greater reliabilty and less downtime) while larger plants get more 'managerial' benefits (better management control, less capital in work-in-progress, less waste, etc).

In the plants with robots installed for production, there are three in which they have made operations more profitable for every one in which they have not. There are 27 users who consider their use of robots to have been worthwhile for every one who does not.

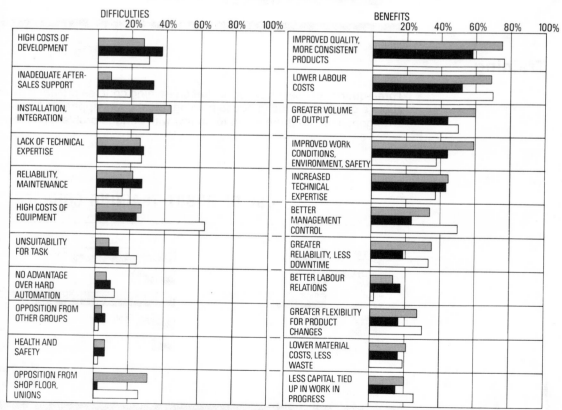

DIFFICULTIES AND BENEFITS FROM ROBOTS

Summary

ROBOT DOWNTIME

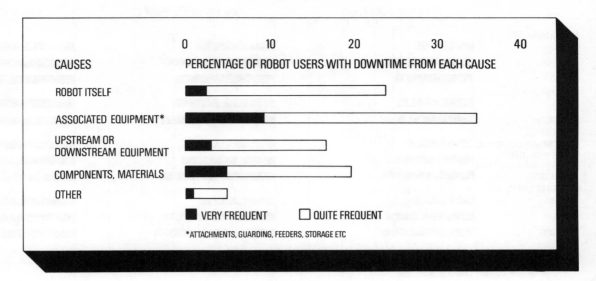

WORTHWHILENESS AND PROFITABILITY OF ROBOTS

FACTORS IN SUCCESS

(Tables 130-133)

The smaller plants tend to do worse than average in their use of robots. Plants in the vehicles industry and plants owned by overseas companies have done rather better than the average, and plants working two or three shifts have done markedly better than those working only one.

Users adopting the turnkey approach have been slightly more successful than the average, in terms of worthwhileness and profitability; and so also have those with the less expensive robots.

The plants which have the most robots and which have used them longest are the ones which have obtained the most benefits from them and found them the most worthwhile.

The plants using robots on the more common applications have tended to do better than those with the rarer ones. Plants which have run into the main difficulties have been less successful than the average, but the majority even of these plants have found that the use of robots has been worthwhile and profitable.

Summary

FACTORS IN SUCCESS: PLANT CHARACTERISTICS

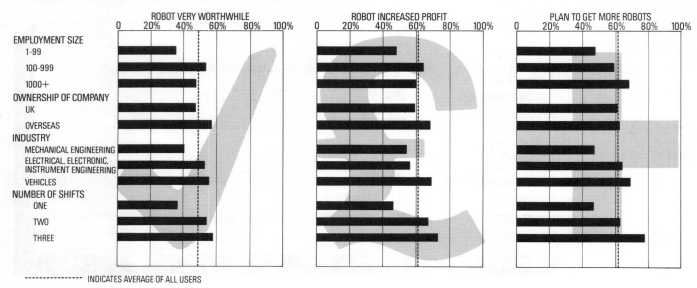

---------------- INDICATES AVERAGE OF ALL USERS

FACTORS IN SUCCESS: APPROACH TO INTRODUCTION OF ROBOTS

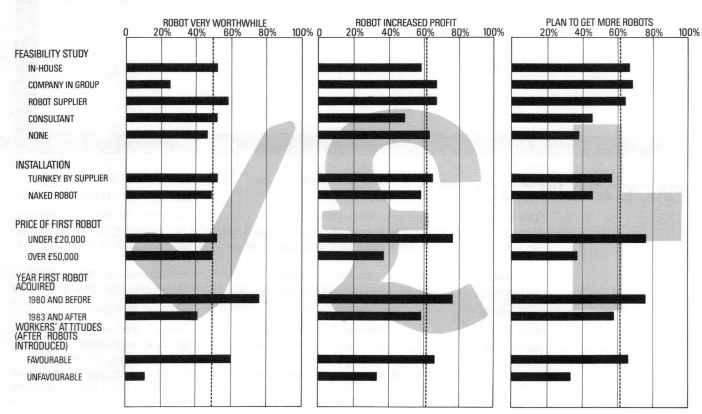

---------------- INDICATES AVERAGE OF ALL USERS

Summary

FACTORS IN SUCCESS: NUMBER AND APPLICATION OF ROBOTS

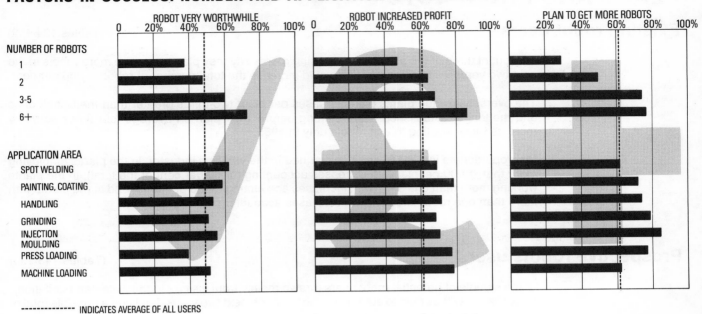

·············· INDICATES AVERAGE OF ALL USERS

FACTORS IN SUCCESS: DIFFICULTIES AND BENEFITS EXPERIENCED

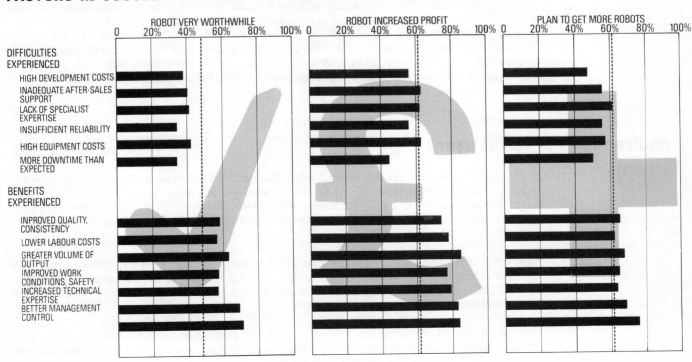

·············· INDICATES AVERAGE OF ALL USERS

Summary

PLANS AND PROSPECTS

Existing robot users
(Tables 134-142)

Two out of three of the plants already using robots say they plan to acquire more robots in the next two years — on a scale which would increase the total number of robots by 60 per cent.

The overseas-owned plants plan increases per plant two-thirds greater than the British ones and the longer established users starting before 1981 plan on average four times as many more as those starting to use robots only in 1985.

Disproportionate increases are also planned in the vehicles industry; in the plants employing more than 1,000 people (with increases accounting for 60 per cent of the total); in the plants working more than one shift (with increases accounting for 80 per cent); and in the plants with more than one robot already (with increases accounting for 80 per cent).

Prospective robots users
(Tables 143-150)

Of the sample of potential robot users (drawn mainly from visitors to the Automan exhibition), 15 per cent say they plan to start using robots in the next two years, implying a possible total of about 800 new users.

They expect between them to acquire more than 2,000 robots. As with existing users, the additional robots are expected to be mainly in the larger plants and in the multi-shift ones.

Growth prospects
(Tables 151, 152)

If the plans of existing users and prospective users are realised, it can be calculated that the total number of robot users in Britain will more than double in two years and the total number of robots in use will go up by 135%. Alternative assumptions, however, suggest an increase over two years of only one fifth in the number of users and only one third in the total number of robots.

The high assumptions are consistent with the intentions declared by the firms in the surveys, the percentage rates of increase of users experienced in Britain in the early 1980s and the absolute rate of increase experienced in Germany in 1985. The low assumptions are consistent with the lower sales reported recently by robot suppliers and the shortfall of recent performance below earlier expectations. Whether the actual outcome turns out to be nearer the higher figures or the lower ones will depend partly on how far the requirements of the users are met.

Requirements of robot users
(Tables 153-155)

In the view of the firms already using robots, the most important obstacles to further use are financial and economic ones. 70 per cent say that cheaper robots would be particularly helpful, 53 per cent ask for cheaper associated equipment, 50 per cent for more government support, 41 per cent for an upturn in the economy and 21 per cent for easier finance for investment.

Next most important are problems related to technical expertise — 56 per cent of users look for easier programming, 43 per cent for less need for special skills, 39 per cent for easier maintenance, 35 per cent for greater reliability and 26 per cent for better after-sales service from suppliers.

The third group of problems are concerned with the performance of the robots themselves. Many users desire improvements in sensors, speed, accuracy, versatility or payload, but are not necessarily ready to pay more to get them.

Summary

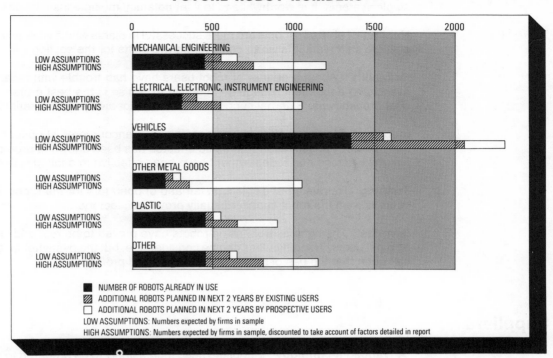

POLICY IMPLICATIONS

Two general points stand out and provide the context within which specific action by users, suppliers and government needs to be considered.

The first is that the use of robots has been successful: 81 per cent of users see their use of robots as having been worthwhile; 61 per cent say robots have made their operations more profitable; and 61 per cent plan to follow up their words with action by buying more robots. This is the basic evidence for backing robots.

The second point is that, if robots have been successful, it is not as isolated pieces of equipment, but as a key element in wider production systems using advanced manufacturing technology. It is in that context that they need to be offered by suppliers, adopted by users and supported by government.

Robot users

Awareness Prospective users need to draw fully on all the sources of information and advice available to them, so they can see clearly the benefits to be won and the difficulties to be overcome.

Feasibility Feasibility studies are worth the trouble — half of them recommend something other than robots. Firms with the necessary expertise do their own studies. Others mostly find the robot supplier satisfactory for this.

Installation Installation and integration need to be planned properly — they give difficulties to one user in three, and the robot cannot earn money until they are overcome. Most first-time users take the prudent course of buying a complete 'turnkey' package from a supplier. Some install robots themselves because they know what they are doing. Others do it just to save money. They usually pay more in the end.

Payback First-time new applications rarely achieve the two year payback commonly demanded but they mostly pay off well over a longer period. Appraisal systems need to take account of this and of indirect benefits in product quality, labour relations, safety and company image.

Summary

Number of robots *Groups* of robots are more likely to pay off — about twice as likely as single systems. It is worthwhile to look at potential multiple installations.

Number of shifts Robots are more successful in plants which work two or three shifts, and in a single shift plant it is worth considering two shifts for the section with robots.

Reliability About a quarter of robot users have had trouble with reliability, most often with associated equipment. Adequate in-house expertise is the best protection against problems over breakdowns. External support services do not always prove satisfactory.

Industrial relations Very few companies have encountered shopfloor opposition to robots. Full and early consultation with those affected has been shown to promote positive attitudes. Natural wastage and redeployment have been effective in dealing with displaced jobs.

Training Retraining of displaced staff and of operators for the robots is essential. Training courses from the robot supplier usually prove satisfactory.

Benefits The benefits from using robots have proved to be real and important and enjoyed by most users. The difficulties can be considerable, but the overwhelming experience of users has been that the use of robots is worthwhile and profitable.

Suppliers

Advanced technology Suppliers need increasingly to be prepared to present robots as part of wider advanced manufacturing systems.

Turnkey installations These will probably need to become still more widely used as use of robots spreads to smaller firms without the expertise to handle installation on their own.

After-sales support Inadequate back-up from suppliers is seen by users as one of the biggest difficulties at present. It will be important to train more key technical staff to improve the service where it falls short of needs.

Reliability Insufficient reliability is still a major worry to users. There will be market gains for suppliers who build up a reputation for reliable equipment and thorough planning of integration.

Costs High costs of equipment are a major deterrent to prospective users and there should be big rewards for any supplier who can bring them down.

Ease of use Many users have only limited technical expertise and design improvements to make robots easier to install, program, maintain and repair are much sought after.

Performance Users are also looking for better sensors, higher speed, greater accuracy, more intelligence, greater versatility and more sophistication generally — but, for the most part, seem reluctant to pay much more in order to get them.

Government

General economic situation The recession is seen by suppliers and users alike as a major barrier to the adoption of advanced manufacturing technology. Suppliers associate more buoyant economies in other countries with brisker robot take-up. 41 per cent of robot users consider an upturn in the economy would be particularly useful in enabling them to make effective use of robots in the future.

Interest rates The present high interest rates for industrial investment are seen as putting British industry at a significant disadvantage relative to overseas competitors. They are a major factor in financial controllers' insistence on short payback periods which effectively rule out the use of robots.

Summary

Investment support If general interest rates cannot be brought down, there is a strong case for selective support for investment in advanced manufacturing. The recent support scheme made a crucial impact on robot investment decisions and its removal may have a drastic effect on an already flattering rate of take-up. Support could take the form of interest relief or tax allowances, but cash grants are probably the most effective way of helping the small and medium firms. A rate of 33.3 per cent, as in the original support scheme, would seem appropriate.

Awareness and demonstration An expansion of the AMT information programme would be helpful in providing advice for industry and visits to 'demonstration firms'. More publicity would make it more widely known and effective and regional AMT centres would make the service accessible throughout the country.

Feasibility study grants These have proved useful and should be continued. Some of the studies undertaken by consultants, however, seem to have had disappointing results, and it may be that more stringent vetting is needed.

Education The shortage of production engineers with expertise in advanced manufacturing technology is a crucial point of weakness. More are urgently needed and provision of relevant courses in higher education needs to be expanded. Consideration also needs to be given to the education and training of the 16-18 age group if we are to have the multi-skilled flexible workforce demanded by the use of AMT.

There is near-unanimity in industry that robots and AMT generally form a special case needing special treatment. While most of the action has to come from the robot users and suppliers in industry itself, there are also a number of things that government can and must do to enable them to get on with the job.

1 INTRODUCTION

Industry must make full and effective use of new technology, particularly in its more advanced forms, if it is to be competitive and successful, indeed, if it is to survive at all. In recent years there has been increasing concern lest Britain's industry may be falling behind its overseas competitors in the speed and enterprise with which it adopts the more advanced kinds of factory automation such as robots.

Figures collected by the British Robot Association show that it is not only the United States and Japan which use more robots than Britain. West Germany, with an industrial base of similar size, has two and a half times as many robots in use, and in each of the past four years the increase in the number in use there has been twice as great as in Britain. The *increase* in the number of robots in West Germany in the one year 1985 was about as great as the *total* number of robots installed in Britain in all the previous years combined. It appears that France also has more robots in use than Britain, and so also has Italy, while Sweden has only slightly fewer than Britain despite its much smaller industrial base.

Equipment makers and consultants have claimed that in Britain robots are not being used in industries and activities where there is apparent scope. Furthermore, many of the existing robot users are not using them to the full. The government has established support schemes to encourage the use of robots and also other kinds of advanced manufacturing systems, but take-up appears on the whole to have been less than was hoped.

In view of this the British Robot Association took the initiative in proposing that there should be a study to find out what has been holding things back and what can be done by industry and government to speed up the adoption of this key technology. Their proposal led to the provision of funding by the Gatsby Charitable Foundation, the Department of Trade and Industry and the National Economic Development Office to enable PSI to undertake a study of robots in British industry. The aim of the study has been to establish the facts about the current use of robots, to identify the issues which influence the effectiveness of their use, and to formulate proposals for action by industry and policy initiatives from government to ensure fuller exploitation of the opportunities available.

It needs to be remembered, of course, that greater use of robots is not something which is necessarily desirable in all situations. Although for many tasks they are a highly efficient production tool capable of performing operations with greater speed and consistency than can be achieved by other means, and they have the special advantage of flexibility in that they can be reprogrammed for new tasks (for example when there is a change in model or assembly method), there are some less complex tasks where simpler pick-and-place machines may be a more economical solution. And there are some where the robot's flexibility is not needed and automatic equipment dedicated to a particular job may be more appropriate. There are others again where non-automated equipment is still the most cost-effective solution. And there are also others where human operators can still perform to better effect, or at lower costs, than robots. Finally, it needs to be stressed that robots and other kinds of advanced equipment are no more than parts of a total production system, and it is the efficiency of the system as a whole, rather than of any of its parts separately, which is what matters.

Nevertheless, it was felt appropriate to focus the study on robots, partly because of their importance in their own right, partly because they can be used as an indicator of the extent of adoption of advanced technology more generally, and partly because they can readily be de-

Introduction

fined, identified and counted and so are more amenable to general measurement and analysis than the diffuse spectrum of systems of which they form a part. For a study aiming to quantify what is happening on a national scale it therefore seemed that the most appropriate course was to concentrate initially on industrial robots.

The study

The study took the form of two major postal surveys, one of existing users of robots and the other of non-users with potential to become users in the future, and a complementary programme of case-study interviews. The postal surveys were undertaken concurrently at the end of 1985 and the beginning of 1986 and the follow-up interviews in the spring of 1986.

There is no list available of all the robot users in Britain, and the sample for the users survey was therefore built up from a number of different sources: firms which had received support for robots from the Department of Trade and Industry; plants which had been shown to be users in the 1985 PSI survey of microelectronic appplications in industry; plants identified as users by robot suppliers; company members of the British Robot Association; and firms which had sent representatives to the Automan exhibition at the 1985 Advanced Manufacturing Summit. The response rate was very high for a survey of this kind — 57 per cent of the known users approached completed the full questionnaire and a further 16 per cent completed an abbreviated version covering just a few of the key questions. Altogether the effective sample amounted to 248 plants, probably about one third of all the robot users in Britain, plus a further 78 who completed only the abbreviated questionnaire. In plant size and industry composition the sample matches very closely the national profile of robot users obtained in the 1985 PSI industry survey, and it seems likely that the sample represents a true cross section of all the robot users in Britain.

The subjects covered by the survey included: the characteristics of the robot users; the ways they set about introducing their first robot; the number of robots in use and the applications on which they are employed; the attitudes of the workers directly affected; the effects on employment and the requirements for training; the difficulties and benefits expected and experienced; the degree of success achieved and the plans for the future. In addition, the firms which had received government support grants had a special questionnaire with supplementary questions about these.

The sample for the survey of non-users was based mainly on visitors to the Automan exhibition, selected to exclude those who were not potential users of robots. They were supplemented by plants who in the 1985 PSI industry survey had said they expected to be using robots in two years time and by company members of the British Robot Association. In the analysis they were divided into three groups: those who had plans to use robots within two years, those who had the use of robots 'under consideration' and those who had no definite plans at the time.

The non-users completed a shorter questionnaire which concentrated mainly on their future plans and expectations.

Details of the samples and other particulars of the surveys are given in Appendix I and in Tables A–E at the back of it. The questionnaires used are given in Appendix II.

The third part of the study consisted of case-studies of selected robot users and prospective users to elicit mainly qualitative information about attitudes, aspirations and the reasons underlying behaviour, and to explore further aspects too subtle or sensitive to be covered adequately in postal surveys.

It is believed that the two postal surveys, together with the interviews, constitute the largest exercise of its kind yet undertaken in Britain and provide the fullest and most broadly based information about the use of robots in Britain and the factors which affect future prospects.

The report

Chapter 2 considers what robots are and what place they have in advanced manufacturing technology, provides estimates of the total number of robot users and the total number of robots in use and maps out the extent of their diffusion throughout industry.

Introduction

Chapter 3 describes the main characteristics of robot users, particularly in terms of plant size, industry and shiftwork patterns.

Chapter 4 examines the processes involved when plants introduce robots for the first time: where the initial idea of using robots originates, who takes the decision to go ahead, who undertakes the feasibility study, what approach is taken to installation and how much time is needed to get the robot into production.

Chapter 5 looks at the government support provided to help firms introduce robots. These took the form of grants towards the costs of feasibility studies, purchase of robots and associated equipment and work on developing applications for robots. It identifies how far different kinds of firm are aware of the support available, the reasons why some firms do not apply for support and the effects of the support on the actions of the firms which get it.

Chapter 6 sets out how the robots are deployed, the number of robots per plant, the distribution of robots between plants of different sizes and in different industries, and the number of hours a week in which they are put to work.

Chapter 7 sets out the different kinds of application on which robots are employed by existing users and the kinds of application envisaged by the plants contemplating using robots in the future.

Chapter 8 examines the impact of robots on people in industry: the attitudes of the workers directly affected and the differences in attitude before and after the introduction of robots; the extent of consultation with shopfloor workers and negotiation with trade unions and the effect of these on their attitudes; the impact of the introduction of robots on employment; the extent to which opposition from shopfloor workers, unions or other groups is a factor affecting the introduction of robots; the numbers of people sent on training courses with the various different organisations providing them, and the extent to which the firms using robots regard the present training arrangements as satisfactory.

Chapter 9 identifies the main difficulties expected by potential users and by existing users before they introduced robots and compares these with the difficulties which have been experienced in practice: the high costs of the robots themselves and of work on applications development; difficulties with installation initially and with product and system reliability subsequently; lack of in-house specialist technical expertise; and inadequate after-sales support from suppliers.

Chapter 10 goes on to consider the many important benefits found by the plants using robots, for example in improved quality of product, lower labour costs, greater volume of output, improved work conditions, increased technical expertise and better management control. It goes on to demonstrate how the extent to which they are experienced varies among different industries and plants of different sizes and increases with shift working, with more years of use of robots and with larger numbers of robots in an installation. Finally, it shows the extent to which the firms using robots have found them to be worthwhile and to have made their operations more profitable.

Chapter 11 analyses the various factors which have been associated with success in the use of robots in the experience of the firms already using them.

Chapter 12 brings together the plans of existing users and potential users in order to map out the changes in prospect in the numbers of robots and robot users and how they will be distributed. It goes on to set out the requirements of users and prospective users and explains how the degree of success achieved in meeting these requirements will affect the extent to which robots are in fact used in the future.

Finally, Chapter 13 sets out the implications of the facts revealed by the study and the analysis made of them in order to make clear the action which needs to be taken by robot users, robot suppliers and the government.

The set of more than 150 reference tables gives the full figures of the use of robots in Britain revealed by the postal surveys. Appendix I gives particulars of the survey and other aspects of the study. Appendix II reproduces the questionnaires used in the postal surveys. Appendix III summarises the help provided by the Department of Industry's support scheme for robots. Appendix IV gives a glossary of selected technical terms commonly employed in discussions of manufacturing automation. Appendices V and VI give details of other research carried out by PSI on the impact of new technology.

2 EXTENT OF USE

Defining robots

Before considering the extent to which robots are being used it may be as well briefly to consider what a robot is. For the purpose of the study we have used the definition of the British Robot Association (BRA):

> An industrial robot is a reprogrammable device designed to both manipulate and transport parts, tools or specialised manufacturing implements through variable programmed motions for the performance of specific manufacturing tasks.

Or, as the Department of Trade and Industry puts it more briefly:

> A robot is a reprogrammable mechanical manipulator.

This might seem straightforward enough, but in practice things are less clear cut. In some other countries, notably Japan and the Soviet Union, definitions are used which include many machines which in this country would be regarded not as true robots but as the simpler pick-and-place machines. And even in Britain the definition is not always easy to apply. There are 'modular' robots which, although they are, strictly speaking, reprogrammable, have only a limited versatility and in practice are often virtually dedicated to a single type of application. On the other hand, there are 'reprogrammable' pick-and-place machines and other kinds of 'hard automation' which offer considerable flexibility and can in practice be modified for quite substantial changes in applications.

One characteristic of robots in which they differ from much other reprogrammable production equipment, such as the machines used for cutting steel plate in the shipbuilding industry and those used for cutting cloth in the clothing industry, is that they can *manipulate* the part that is being worked on. However, in many robot applications, for example in spotwelding in the motor industry, the *part* is manipulated by the assembly line or other equipment, and it is only the *tool* which is manipulated by the robot.

Another distinctive characteristic of robots is that one machine can perform a *succession* of tasks, in contrast to the usual pattern in, for example, the cotton industry, where a succession of *separate* programmable machines are linked to one another by transfer equipment. However, the integration of successive stages in a single machine is not invariably an advantage, particularly if the flows of work come through the successive stages at different or irregular speeds; and anyway, in some cases the successive stages can be, and are, incorporated into a single fixed machine.

Perhaps the most important characteristic which differentiates robots from other kinds of automated production equipment is their *versatility*; the same machine can be redeployed to new tasks of different kinds to meet changing needs. It should be noted, however, that even this characteristic is often not exploited. In practice it is common, once a suitable application has been developed, for a robot to remain in the same use for long periods and to be treated very similarly to equipment dedicated to the particular task.

Thus the boundary line is becoming increasingly blurred and there is some dispute, not only

Extent of Use

over whether a particular machine should be regarded as a robot or not (particularly in injection moulding applications), but even as to whether it is meaningful and useful to seek to draw a distinction between robots and other kinds of advanced manufacturing equipment at all.

In a broad study of this kind it is not practicable to investigate and adjudicate on each piece of equipment, and it should therefore be noted that, while the attention of respondents was drawn prominently to the BRA definition, the exact figures they have given us will inevitably have been influenced by the ways they chose to interpret the definition. A more stringent, or less stringent, interpretation, or a different definition, would result in rather different figures.

Total number of robot users and robots

Quite apart from difficulties over definition, there is no certainty about the exact number of factories currently using robots in Britain. The survey results, however, enable us to make a reasonably reliable estimate of the total. The PSI industry survey suggested that in early 1985 there were in the region of 560 factories using robots in Britain, and as 24 per cent of the users of robots in the survey have been using robots only since 1985, it may be deduced that by the beginning of 1986 there were a total of about 740 users. Estimates by four other methods (described in Appendix I) based on data from the robots survey and the earlier industry survey suggest figures of 570, 740, 790 and 840 respectively, and the mean of all five methods works out at 730. Accordingly, it seems reasonable to take 740 as a practical working assumption for the total number of robot users at the time of the survey with fair confidence that the true figure is likely to be within the range of 700-800 and may well be very close to the one adopted.

The figure for the total number of robots in the UK at the end of 1985 given by the BRA in its annual publication *Robot Facts* is 3,208. The PSI industry survey provides a basis for alternative estimates both higher and lower than this but, in the absence of strong reasons for preferring any particular alternative figures, and in view of the fact that the BRA figure is widely used, the most practical course is to use this as the working assumption of the total number of robots for the purpose of the calculations in the survey analysis.

The two figures, 740 users with 3,208 robots, imply an average of 4.3 robots per user, which is the figure resulting from the replies of the robot users in the survey. The fact that the three figures fit together suggests that, while none of them may be exactly correct, at any rate none of them is likely to be incorrect by a substantial margin.

Growth in number of robot users and robots (Tables 16-20; Question 14)

In some ways more important than the total number of robots in use at any one time is the growth in numbers over a period of time. Clearly, the number of robot users over a number of years must be a matter of rather more uncertainty than the the number in 1986, but the information from the survey respondents about the year in which they first started using robots provides a basis for reasonably reliable estimates. And figures for the total number of robots for earlier years are provided in the BRA's annual publication *Robot Facts*. The picture that emerges from the BRA information is given in the table below.

Growth in number of robot users and robots in UK

Year	Robot users		Robots *	
	no. at start of year	annual change per cent	no. at start of year	annual change per cent
1981	100		371	
		+42		+92
1982	150		713	
		+50		+62
1983	220		1152	
		+56		+52
1984	350		1753	
		+62		+50
1985	560		2623	
		+31		+22
1986	740		3208	

* Source: BRA *Robot Facts*.

Extent of Use

The percentage rates of increase in the numbers, both of robots and robot users, are very high, leading over the five year period 1981-1986 to a more than seven-fold increase in the number of robot users and a more than eight-fold increase in the number of the robots themselves. However, these increases were from a very small base at the start of 1981 and the total numbers in 1986 are still very small. Moreover, the increases in numbers of both robot users and robots in 1985 were less in absolute terms than in the year before, and much less in percentage terms than in any of the four previous years.

Extent of diffusion

(Tables I-6)

The small size of the robot base in Britain can be measured in terms of the modest extent of diffusion in industry as a whole. The PSI survey of microelectronics in industry in early 1985 found that of all the factories in the country employing more than 20 people, about half were using microelectronics technology in their production processes in one way or another and about one in seven was using CNC machine tools. However, less than one factory in 60 was using robots, and only one in 30 the simpler pick-and-place machines.

The same survey found that nearly one factory in 20 was expecting to be using robots two years later, by early 1987. This expectation seems almost certainly destined to be proved much too optimistic, but even if it were fulfilled it would still mean that the other 19 factories out of 20 would *not* be using robots. It seems reasonable to conclude, particularly in the light of developments in other countries such as Japan, that the rates already achieved and immediately envisaged fall well short of saturation level.

As might be expected with such a small, select group, the robot users are to be found predominantly in the more sophisticated parts of British industry. Robots tend to be used in plants which are already users of other applications of new technology: the robot-using plants are three times as likely as the average to be using Computer Aided Design, automated testing and quality control and computer-integrated control of machine groups, five times as likely as the average to be using automated handling systems and seven times as likely as the average to be using automated storage systems.

Robot users are more common in the vehicle, aircraft and electrical, electronic and instrument engineering industries than in others, and much more common in the larger plants than in the smaller ones. One in four of plants employing more than a thousand people uses robots, compared with less than one in a hundred of those employing less than 200 people. And, ominously, they are three times as common in overseas-owned groups of companies as in British owned ones.

Larger companies seem from the evidence of the survey and the interviews to be more likely to see *learning* about the technology as one of the main reasons for using robots. These firms also tend to have greater interest and expertise in planning for integration of manufacturing processes and the use of other forms of advanced technology. Larger companies tend to have the financial resources and the in-house expertise to make greater investments in technology than small firms, and to experiment more instead of needing to put robots straight into production. All of these factors are clearly at work where overseas-owned groups are concerned, and there may be a more dynamic attitude in the parent company, and opportunity to draw on technology and management systems developed by it, which have an impact in the UK. Among the companies interviewed which were US-owned there was a positive attitude to new technology and considerable resources went into monitoring the latest developments in advanced manufacturing technology in general. One interviewee claimed that his firm's takeover by an American group had led to a more positive attitude among management towards investment in new technology.

3 CHARACTERISTICS OF ROBOT USERS

Size
(Tables 7, 8; Question 3)

The most striking characteristic of robot users is that they tend to be large. Altogether three-quarters of them are plants employing more than 200 people and one third of them are plants employing more than 1,000 people, while fewer than one in eight employ less than 100.

Employment size of plants using robots

row percentages

1-99	100-199	200-499	500-999	1000-	TOTAL
12	11	27	17	32	100

They also tend to be large in other ways. Only one in twenty has an annual company turnover of less than £1 million and one in four has a turnover of more than £100 million. Fewer than one third of them are in single plant companies while more than one third are in companies with five plants or more. The large plants also tend to have greater numbers of robots so that they account for an even higher proportion of the total number of robots in use.

There are a number of reasons for this preponderance of large users. First, they more often have a scale, range and sophistication of production operations to give good scope for the use of robots; secondly, they more often have the specialist expertise to be able to use them effectively; thirdly, they have the money to afford to buy them; and finally, they tend to be more aware of the opportunities offered by robots.

There are signs, however, that the smaller plants are beginning to become a little less sparsely represented. Ninety per cent of the smallest robot users employing less than 100 people started using robots only in the last three years, compared with 65 per cent of those employing over 1,000 people. It seems likely that there were none of these small plants at all using robots before 1981, but they accounted for 20 per cent of all those starting to use robots in 1984 and 14 per cent of those starting in 1985. In contrast, the largest plants employing over 1000 people accounted for 39 per cent of those using robots before 1981, but for only 29 per cent of those starting to use them in 1984 and 25 per cent of those starting in 1985. However, there is still a very long way to go before there is more than the most marginal shift in the tendency of robots to be used predominantly in the larger plants.

Industry
(Tables 10, 11, 13; Questions 6, 7)

About 60 per cent of all robot users are in the vehicles and engineering sectors, 13 per cent in other metal goods and 9 per cent in plastics, with the remainder scattered thinly between a wide range of other industries. The robot users in vehicles and aircraft tend to be particularly large in terms of plant size, number of plants and company turnover, while the users in mechanical engineering and plastics are quite often small.

Characteristics of Robot Users

Industry sector of plants using robots

row percentages

Mechanical engineering	electrical, electronic engineering	vehicles, aircraft	other metal goods	plastic products	other	TOTAL
23	15	22	13	9	18	100

To some extent the distribution of robot users among industries reflects differing potential for use in different kinds of activity, but it is also a matter of awareness of the scope that exists and there are some signs that the mix of industries using robots is tending to widen. It is mainly in the past three years that robots have been used in industries such as chemicals, paper and ceramics, and 62 per cent of the users in mechanical engineering started only in the last two years. The proportion of robot users accounted for by the mechanical engineering industry has risen from 15 per cent of those starting before 1981 to 23 per cent of those starting in 1985, while the proportion accounted for by the vehicles industry has fallen from 30 per cent of those starting before 1981 to only 14 per cent of those starting in 1985.

A significant proportion of robot users, about one in nine, are in companies which themselves make robots or associated equipment, although not necessarily at the same plant. No doubt these companies are much more likely than most to be aware of the advantages of robots and to have the specialist expertise to use them effectively. A higher than average proportion of these users have robots in experimental development work to help them to meet the needs of their customers, and those using them in commercial production presumably find it a help in selling to be able to demonstrate robots operating successfully in their own plants.

Shiftwork

(Table 12; Question 4)

A third striking characteristic of robot users is the prevalence of shift work. Robots tend to be expensive, but they are capable of working round the clock. So, all else being equal, they are more likely to be cost-effective if used in a shop which runs for more than a single shift. Hence it is not by chance that only 29 per cent of the plants with robots installed for commercial production work only a single shift in the sections where robots are used. Forty-five per cent run a two shift system and 22 per cent run three shifts. And as the plants with more than one shift have on average about three times as many robots as the single shift ones, they account for about five out of six of all the robots in use.

There are signs, however, that the use of robots in single shift plants is increasing. Over half the users with their robots still in the pre-production stage at the time of the survey were in single shift plants, and the proportion of all robot users accounted for by the single shift plants rose from 21 per cent of the plants using robots before 1981 to 40 per cent of the ones starting to use robots for the first time only in 1985.

4 INTRODUCTION OF ROBOTS

Source of initiative
(Tables 21-25; Question 9)

The source of the initiative to use robots varies greatly from company to company, reflecting the great diversity in forms of company organisation. Within any one company there are often also differences between the site of the original idea to consider using robots and the location of the final decision to go ahead.

Origin of idea and place of decision to use robots

row percentages of robot users

	head office	company board	plant management	department	NA	TOTAL
Idea	7	27	42	21	4	100
Decision	15	59	17	3	5	100

With more than two-fifths of all robot users the source of the original idea of using robots is at plant management level, but with one third of the users the idea originated at company board or head office level and with a further fifth at department level. In the small plants employing under 100 people the idea is much more likely to originate at board level than it is in the large ones employing more than 1,000 people (68 per cent against 10 per cent), whereas in many of the largest ones the original idea arises at department level, but in none of the smallest ones. These differences, which were evident in the interviews no less than in the survey figures, reflect the consideration that in small plants the company board is often close to the action and embodies much of the company's technical expertise, while in very large plants the company board may be in a position to deal only with broad strategic decisions. It may therefore feel it necessary to devolve more initiative to plant management, or even departments, which in a very large plant may have considerable expertise and resources.

There are also differences between industries, with the original idea more often coming from the company board in mechanical engineering, or more often from plant or departmental management in the vehicles industry, but to a large extent these differences reflect variations between industries in average size of plant.

In all kinds of organisation the final decision tends to be taken at a higher level than the source of the original idea, with no less than three-quarters of all robot users taking the final decision at company board level or head office. Here too, however, there is a tendency for greater devolution with larger plant size. 84 per cent of the smallest plants take the final decision at company board level, and only three per cent leave the final decision to the plant management. Of the largest plants, only 47 per cent take the final decision at company board level, but 27 per cent leave the final decision to plant management.

There are also differences between industries, with more readiness to devolve decisions to plant management or below in vehicles and electrical and electronic engineering than in mechanical engineering or other metal goods, where plant sizes tend to be smaller and plant managements customarily are accorded less autonomy. Also the plants with more than six robots are more inclined to leave the decision to plant or department management than are

Introduction of Robots

those with only a single robot (35 per cent against 13 per cent), partly, perhaps, because of greater familiarity with robots and partly, perhaps, because these tend to be the larger plants anyway.

The motivation for considering the use of robots varies widely, but several factors stand out and were emphasised in the interviews. Improvements in output, quality and consistency, the prospect of lower labour costs, improved conditions of work and increased expertise in using new technology are all significant factors. (See chapter 10 for more details about the benefits expected by users and prospective users.) It is also clear from the interviews that competitors' plans for, or use of, robots is often an incentive to management to investigate the potential of robot systems. For some companies there has been pressure from customers for robots to be used in the manufacture of particular products.

Feasibility studies

(Tables 26-29, 31; Questions 10,II,15)

It is usual to undertake a feasibility study before introducing robots, but one in ten of present users went ahead without one. This could sometimes be due to confidence in the firm's expertise and competence to introduce robots successfully, but it could also be due sometimes to unawareness of the potential problems involved. In the case of the smaller plants it could reflect more informal decision-taking arrangements and the absence of a need to make a formal study as part of the process of securing approval from head office. Whatever their reason, the plants which did not undertake a feasibility study took half as long again as the others to get their first robot into commercial production, 20 per cent of them needing more than twelve months for this. Their responses to the introduction of robots do not, however, appear to be very different from those of plants which did carry out feasibility studies, in that in the end about half of them have found their use of robots very worthwhile and nearly two thirds of them have found the robots have made their operations more profitable.

The most frequently adopted course, followed by more than one third of all the robot users, has been to undertake their own feasibility study in-house. This has been a particularly common approach with the larger plants, which more often have the technical expertise to make this practicable, and also with the users of larger numbers of robots — hence nearly half of all the robots are used by plants which proceeded in this way.

Organisation undertaking feasibility study

row percentages of robot users

Own company in-house	other company in group	robot supplier	consultant	no study	NA	TOTAL
38	7	30	14	10	4	103*

* Total adds to more than 100 because a few plants had feasibility studies undertaken by more than one organisation.

The next most common course, followed by rather less than one third of all the users, has been to have a feasibility study undertaken by the supplier of the robot. While this has involved getting an appraisal from a source that is not disinterested, this has been felt to be more than offset by the advantage of being able to draw on the supplier's specifically relevant expertise and familiarity with the particular equipment envisaged. The plants which have followed this course have not got into production any more quickly than the others, and nor have they ended up in buying more robots than the average, but they have fared somewhat better than the average, as shown by the proportion finding that their use of robots has been very worthwhile and has made their operations more profitable.

A few plants have had a feasibility study carried out by another company in the same group and 14 per cent of all users have used an outside consultant. The former group has tended to be small plants using only one or a few robots. On average they have managed to get into commercial production a little more quickly than the others, but are less inclined than the others to acquire more robots and include a well above average proportion of those who have not found their use of robots profitable. Those using the consultants express only a moderate degree of enthusiasm for them, with 39 per cent rating the consultant's report good and 39 per cent only fair, and 7 per cent rating it excellent, compared with twice this percentage rating it poor or useless. While some of the consultants are highly praised by their clients, others are the object of vigorous complaint. Several interviewees express scepticism about the degree of technical expertise possessed by some consultants, and say they thought that they would have done just as good a job as the consultants they used. They feel consultants did not come up with anything original and had simply 'recycled' information already available to the customers.

Feasibility studies have also been undertaken by plants which have not become robot users. Of the non-users in the sample, 11 per cent had a feasibility study planned, 4 per cent had one currently in progress, and 21 per cent had one already completed. However, with over half the last the feasibility study recommended against the use of robots. Where the study recommended in favour of using robots, half the plants said they expected to acquire robots within the next two years, and the other half said that they had it 'under consideration'. Where the study recommended against the use of robots, only 7 per cent said they had plans to use robots within the next two years, and where no feasibility study was even planned only 4 per cent said they planned nonetheless to use robots in the next two years.

Approach to installation

(Tables 30-32; Question 16)

Difficulty arising from the installation of the first robot and its integration into the production process is the problem with robots that is most widely feared in advance of adoption, and is also among the difficulties most frequently experienced subsequently. The approach adopted to installation of the first robot is therefore a matter of considerable importance.

The approach followed by nearly half the robot users has been to have the supplier of the robot provide a complete 'turnkey' package. This is more costly at the outset but enables the new user to benefit from the experience of the supplier and reduces the risks of things going wrong. With subsequent robots used in the same application these plants have often felt able to manage on their own, but for different applications involving different problems they have often found it advantageous to resort to a turnkey solution. The turnkey approach is particularly common in the vehicles and metal goods sectors, with overseas owned companies, and with those whose feasibility study was undertaken by the robot supplier.

Approach to installation

row percentages of robot users

Turnkey by robot supplier	turnkey by consultant	naked robot	other	NA	TOTAL
48	4	38	9	2	100

The next most common approach is for the prospective user simply to buy the robot and make internal arrangements for installing it and getting it into commercial production. This costs less initially, but can give rise to difficulties if the firm lacks the specialist expertise to get the robot running or makes mistakes in planning. Several interviewees regret not adopting a turnkey package for the installation of their first robot. In some cases the robot has been under-utilised because of a lack of programming skills or because unsuitable applications were chosen. For others development costs (such as tooling and re-design of components) and the pitfalls of installation were seriously under-estimated, and the 'DIY' approach did not save money in the end.

This approach has been followed by more than one third of all the robot users in the sample, but by higher proportions of the smaller plants, which are often concerned to keep initial cash outlays to a minimum; by plants which did their own feasibility studies, and may thereby have gained confidence in their ability to handle installation; and by companies in electrical and electronic engineering, who may have been more confident than firms in other industries that they had the necessary expertise in-house. It is also the usual approach when robots are acquired initially for development and experimental purposes and immediate use in commercial production is not envisaged.

A few new users have adopted other approaches to installation, for example a 'turnkey' provided by the consultant who undertook the feasibility study.

Introduction of Robots

Time needed for installation
(Tables 33-35, Question 15)

Altogether, at the time of the survey, 19 per cent of the robot users in the sample were at the stage at which their first robot was not yet in commercial production, 73 per cent had their first robot already installed for production, and with the remaining 7 per cent their first robot was abandoned, out of use or sold. This last situation was normally not an indication of failure, but rather that their original robot was an early machine which had become out of date, or worn out, or was no longer suited to the company's changing needs. Half of this group felt that their use of robots had been 'very worthwhile', all were still using robots, and the majority were planning to buy more in the future.

Of the 19 per cent whose first robot was not yet in commercial production, some were using the machine for development or experimental purposes, but many were involved in preparing their first robot for production. It is rare for a first robot to be able to go on-line immediately, but nearly half of them have gone into production within three months and nearly three-quarters of them within six months. There can be problems, however, with the minority which take longer, particularly for the one in ten which take more than a year to get into production — an expensive experience for firms looking for a payback period of only two or three years. Subsequent robots, however, are normally much easier and quicker to get into production, particularly if they are used on the same application as the first one. One interviewee says, for example, that the first installation took some 15 months from feasibility study to the production stage, but that subsequent installations for the same application have taken much less: only about a month in some cases. Where complex new applications are concerned, however, even an experienced robot user can find that installation time is as long as or even longer than that for the first robot.

Months to get first robot into production

row percentages of robot users

0	1-3	4-6	7-9	10-12	12-	TOTAL	mean
9	35	26	6	13	10	100	6.4

The time needed to get the first robot into production tends to be below average in the plastic products industry, where two-thirds of first robots are in production within three months, and above average in vehicles and electrical and electronic engineering, where one in six takes over a year.

In the plants where a feasibility study has been done by a consultant, or by a company in the same group, the average time needed has been well below the average needed for all new robots. However, in those where no feasibility study has been done the average time has been well above the average for all new robots, with nearly half of them taking more than nine months. The overseas-owned companies also have tended to do rather better than the average, with 57 per cent getting their first robot into production within three months and only 5 per cent needing more than a year.

5 GOVERNMENT SUPPORT FOR USING ROBOTS

The Government has sought to encourage firms to introduce robots by offering various kinds of support: in particular, grants towards the costs of studies of feasibility and grants towards the costs of purchase of robots and associated equipment and the costs of development work on applications. Details of Department of Trade and Industry support schemes are given in Appendix III. What impact have they made?

Grants for feasibility studies
(Tables 36-41, 55; Questions 12, NQI0)

The great majority of the robot users in the survey undertook a feasibility study before going ahead with robots, yet only one in four of them in fact received a government grant towards the cost of the study.

Part of the reason for this is that one in seven of the users was unaware of the availability of the feasibility study grants. Those who did not receive grants for developing robot applications were twice as often unaware as those who did receive grants; and there was also a below average level of awareness in the plants which had their feasibility study done by the robot suppliers and by those which did not have a feasibility study done at all. Those whose study was done by a consultant, in contrast, were all aware of the availability of the grants.

Of the firms that were aware of the grants, rather more than half did not apply for them, including two out of three of the plants employing less than 200 people, of the firms in vehicles and other metal goods, of the firms which did not receive grants for developing robot applications, and of the firms which had had a feasibility study undertaken by anyone other than a consultant. Four out of five of those who did not carry out a feasibility study at all also did not apply.

Awareness and use of support for feasibility studies

row percentages

	Not aware support available	aware but did not apply	intend to apply later	applied but rejected	received support	NA	TOTAL
A	17	45	—	7	25	7	100
B	17	43	19	4	13	4	100
C	30	57	7	1	4	1	100

A — Robot users
B — Non-users planning to use robots in next two years
C — All non-users in survey sample

Firms offered many different reasons for not applying, among the more common being that the money was not essential to them, the amount involved was anyway small, and they were not prepared to face possible 'hassle' or delay in order to get it. There were also some who did not want a study done by a consultant.

Only 7 per cent of the robot users in the survey applied for grants for feasibility studies but were not accepted. Those rejected were predominantly very large plants and plants using more than one robot.

One in four of the plants in the sample received grants for feasibility studies, including above average proportions of the firms in plastic products and in electrical and electronic engineering, of the overseas owned companies, and of the firms awarded grants for developing robot applications. The most successful group of all in this respect were the firms which had feasibility studies done by consultants, no less than 80 per cent of which applied for and received government grants.

Of the firms in the sample of non-users, 30 per cent were not aware of the availability of grants for feasibility studies — nearly double the percentage of the users who were not aware of them. Only 8 per cent of the non-users were intending to apply for a grant in the future, and only 4 per cent had actually received one. However, in the case of non-users who were planning to use robots within the next two years, the position was much closer to that of the firms already using robots: only 17 per cent were unaware of the grants and, although only 13 per cent had actually received them (half the proportion of the users which had received them), a further 23 per cent had an application under consideration or were planning to apply in the future.

Grants for purchase of robots and development of applications

(Tables 42-54; Questions 36-39, NQII)

Until June 1986, government grants were available in some cases to help firms with the purchase costs of robots and associated equipment and with the development costs of getting robots into production. (See Appendix III for details of the scheme.)

Slightly less than half the robot users in the sample have received these grants, and those which have not received support were asked why they had not sought it. The reasons given were varied, with no single one outstanding. The main reasons were:

	per cent of users who did not receive robot grants
not eligible for support	26
not aware support available	14
processing of application too slow	12
conditions of support too stringent	10
did not need support	10
level of support too low	4

The public sector companies which were interviewed regretted not being eligible for the robot grants, and some interviewees were concerned lest the processing of applications might have been too slow and that too much time and expense would have been involved in obtaining the grant. However, none of the interviewees who did not seek support is against the policy of providing robot grants, even if some of them have complaints about processing, eligibility or the amount available.

Of the firms which have received grants, three-quarters were already aware of their availability before their feasibility studies and most of the remainder became aware during the course of them. Much the most important source of information for making firms aware of the support available is the trade and technical press, cited by more than half of them. The next most important source is the robot suppliers, mentioned by 24 per cent, and after that a number of less frequently mentioned sources, including exhibitions, trade fairs, conferences and courses, other companies in the same group, the general press, consultants and radio and television.

When asked what they would have done in the absence of a grant, about half the grant-supported firms say they would not have gone ahead with robots at all, a quarter say that their plans would have been delayed and a quarter that they would have been on a smaller scale. One should not read too much significance into these figures, since the firms had already certified in their grant applications that without a grant their projects would not go ahead as quickly, or on such a scale, or at all; and if in reality they had felt that the grant had made no difference to their behaviour, it would be unrealistic to expect them to say as much in writing on a postal survey form.

However, the views expressed by firms in the course of the confidential interviews leave no doubt that the grants have been an important, often decisive, factor in their decision to go ahead with robots. A particularly important factor has been their contribution to making payback possible within the limits set by their financial controllers. They have also provided a useful addition to the cash flow which had often materially improved the outcome of the project once it was started.

While many had criticisms of the operation of the scheme, in particular the trouble involved in preparing applications and the delays involved in their processing, very few question the policy of support and most are emphatic in their belief in the value of the support provided. Typical comments from interviewees are given below.

> 'There's no way we'd have bought a robot without support.'
> 'We could not have gone ahead without a grant — which is what you have to say to the DTI, but it's absolutely true.'

If those who say that in the absence of a grant they would not have gone ahead at all had not received a grant and not gone ahead, it can be calculated that there would be somewhere in the region of 450 fewer robots in use today and more than 100 fewer planned in 1986 and 1987.

When asked whether, given the knowledge and experience now gained, they would now be prepared to undertake a similar project without a grant, only very slightly more say they would than say they would not; but there is a 2 to I majority saying they would in the larger plants and in the vehicles industry, and a 4 to I majority of the plants which have been using robots for more than five years.

With the non-users the level of awareness of grants is somewhat lower. Only a little over half are aware of the availability of grants for the purchase of robots and only about a quarter of them are aware of the grants for associated equipment and for development work to get the robot into production. Among interviewees who are aware of the grant scheme the consensus is that grants would make a major and usually decisive contribution to the cost-justification of robot projects. Lack of capital or severe doubts about justification are frequently mentioned as obstacles to adoption, and grants are generally seen as a way of encouraging non-users — particularly the smaller firms — to take the risk.

6 NUMBER OF ROBOTS

At the time of the survey, at the beginning of 1986, it seems that there were altogether 700-800 plants using robots in British industry, with a total of about 3,200 robots among them. This implies an overall average of about 4.3 robots per plant, but in fact the distribution between plants was far from uniform.

Number of robots per plant

(Tables 58, 59; Question 17)

No less than 43 per cent of the plants using robots have only a single robot and a further 18 per cent have only two robots, but they account, respectively, for only 10 per cent and 9 per cent of all the robots in use. At the other end of the scale, only 8 per cent of the users have more than ten robots each, but they account for 46 per cent of all the robots, and more than a quarter of all robots are in the 3 per cent of user plants with more than 20 robots each.

Number of robots per plant

row percentages of robot users and robots

	1	2	3-5	6-10	11+	NA	TOTAL
Plants	43	18	19	8	8	4	100
Robots	10	9	18	16	47		100

To a large extent these differences can be explained in terms of different lengths of time in which the robots have been used. Of the plants with only a single robot, nearly 30 per cent are still in the experimental, pre-production stage, and 40 per cent of them acquired their first robot only in the previous twelve months.

With increasing experience of use the number of robots per plant tends to rise markedly. Of the plants which acquired their first robot only in 1985, 70 per cent had only one robot at the time of the survey and none had more than ten robots; in contrast, of those which acquired their first robot before 1981, only 12 per cent still had only one robot at the time of the survey and no less than 24 per cent had more than ten robots. On average the latter, more experienced, users have four times as many robots each as the new users.

The average number of robots per plant appears to have remained within the range 4–5 in each of the past five years. The average has not increased because the tendency of existing users to get more robots year by year has been roughly offset by the tendency for new users to start with only one or two robots each. For example, of the non-users expecting to start using robots in the next two years, one third expect to acquire only one robot and a further one third expect to start with only two.

Type of company

(Table 60, Question 17)

The overseas-owned companies are similar to the British ones in the sense that nearly half of them have only one robot. Where they differ is that far fewer of them have only two robots and far more of them have more than six. As a result, the overseas-owned plants have on average nearly twice as many robots as the British ones and account for about 30 per cent of all the robots in use.

Government Support for Using Robots

Size

(Tables 61, 62; Question 17)

As might be expected, the larger plants tend to have more robots than the smaller ones: 30 per cent of those employing more than 1,000 people have only one robot, compared with 65 per cent of those employing fewer than 100; and 23 per cent have more than six robots, compared with none of those employing less than 100. Hence these larger plants employing more than 1,000 people have, on average, three and a half times as many robots per plant as the small ones, and account for about half of all the robots in use.

It is not only the larger *plants*, but also the larger *companies*, which tend to use robots in larger numbers. The robot users in companies with five or more plants have, on average, twice as many robots per plant as the single plant users, and account for nearly half of all the robots in use.

Industry

(Table 63, Question 17)

There also substantial differences between industries. The vehicles industry accounts for nearly half of all the robots in use, largely because of a number of very large users in it. The industry has double the average proportion of plants with more than ten robots each, a reflection largely of the considerable numbers of robots used in spot welding. In plastics the proportion of users with more than 20 robots each is more than double the average, in this case because of the large numbers of machines used in injection moulding.

Hours of use and shifts

(Tables 64-67, Question 17)

One of the advantages of robots is that they can work round the clock and, in practice, the robots used by the plants in the survey are scheduled to work on average for 68 hours a week — far more than the average for human workers. However, just as there are variations between plants in the number of robots per plant, so also there are important differences in the number of hours a week in which they are used.

In general, the number of hours the robots are in use is related to the number of shifts worked in the section concerned. Altogether, 33 per cent of the plants using robots work a single shift, 40 per cent two shifts, and 22 per cent three shifts, and the average number of hours worked per week by the robots in each is 42, 69 and 100 respectively.

About two-thirds of the small plants employing less than 100 people use robots in a single shift, while of the largest ones employing more than 1,000 people, about half use robots on a two shift basis and about a quarter for three shifts. Two shift use is particularly common in the vehicles and other metal goods industries and three shift working in plastics.

Thus by and large where there are more robots per plant they are also worked for longer hours — in the single robot plants they are used only for an average of 56 hours a week, while in the plants with more than ten robots they are used for an average of 95 hours a week.

Hence the impact of the concentration of many of the robots in a relatively small number of plants is compounded by the fact that in these plants each robot is typically worked for many more hours per week than the robots in the single robot installations. The 8 per cent of robot users with more than ten robots each account between them not only for about 46 per cent of all the robots used but also for about 64 per cent of all the hours worked per week by the robots.

7 APPLICATIONS

Existing robot users
(Tables 68-74, Question 119)

The robot application used in the greatest number of plants is arc welding, used by a third of all robot users. Next most common, each used by about one sixth of all robot users, are assembly, machine loading, painting and coating, and handling. Other relatively common applications, each used by 8 per cent of users, are injection moulding, glueing and sealing, spot welding, inspection, press loading, grinding and training.

There are also many other applications, many of them developed only recently, which are used by only very few plants. These include casting, drilling, cleaning, forging, heat treatment, waterjet cutting, laser cutting, woodworking, food processing and textile handling. Finally, some plants use robots solely for research purposes or for educating staff about the potential of robotics.

However, the number of plants using a particular application is only one measure of its importance. Another, and in some ways more significant one, is the total number of robots used in it. Accordingly, the survey respondents with more than one robot were asked how many were used on each application. Unfortunately, this was the one question which many of them did not answer, and the survey therefore does not provide information on the number of robots used in each application (as opposed to the number of plants using robots in the application or the total number of robots on all applications in each plant).

Fortunately, however, figures for the total number of robots in each application are provided in the British Robot Association's *Robot Facts*. By this measure the most important applications are injection moulding and spot welding, in each of which more than 500 robots are employed. This reflects the frequency of large multiple installations of injection moulding robots in the plastics industry and of spot welders in the vehicles industry. While in these two applications the average number of robots in an installation is about ten robots or more, in most other applications, by contrast, the typical installation consists of only one or two robots.

Robot applications

Application	Number of plants	Number of robots*
Injection moulding	60	551
Spot welding	40	511
Arc welding	240	411
Assembly	130	294
Machine loading	120	287
Painting, coating	100	193
Handling	100	130
Press loading	30	74
Glueing, sealing	40	43

* BRA *Robot Facts*

Applications

As might be expected, spot welding robots are used mainly in the larger plants in the vehicles industry with large total robot installations, while the robots used in injection moulding are more often in plants employing 100–500 people and are predominantly in the plastics industry.

Applications in glueing and sealing, painting and coating, assembly, handling, machine loading and packaging are most commonly in the larger plants. Applications in arc welding are particularly common in the mechanical engineering and other metal goods industries, and those in assembly in the electrical and electronic engineering industries, while most kinds of application are well represented in the vehicles industry.

Prospective users

(Table 75, Question 12)

In the plants contemplating using robots for the first time in the course of the next two years, the three applications most frequently envisaged are the same as for the existing users, but in a different order. Of the prospective users, 38 per cent foresee applications in assembly (compared with 18 per cent of existing users), 28 per cent in machine loading (compared with 16 per cent), and 24 per cent in arc welding (compared with 32 per cent of existing users).

Among the other applications envisaged are handling (15 per cent of the prospective users), injection moulding (12 per cent), grinding (10 per cent), painting and coating (9 per cent), press loading (9 per cent) and packaging (9 per cent).

Although there are some differences from existing users, they do not seem to be sufficiently great to bring about any very substantial change in the overall pattern of applications in the course of the next two years.

8 TRAINING, EMPLOYMENT AND INDUSTRIAL RELATIONS

Attitudes of workers affected

(Tables 76-81, 94, 121; Questions 27-29)

One of the difficulties which many managers expected before they introduced robots was opposition from shopfloor workers or trade unions. In the event, their apprehensions have nearly always proved unfounded. Only 6 of the 248 plants in the sample (2 per cent) say they have actually experienced difficulties due to opposition from shopfloor workers or unions. An equally small number report difficulties due to opposition from top management, and three times the number say they have experienced difficulties due to opposition from other groups in the company. In the case-study interviews managers almost invariably considered shopfloor workers and unions to be well aware of the need to modernise plant and introduce new technology where appropriate. Opposition from senior management usually seems to be based on scepticism about the return on investment from robot systems. In a few cases opposition is reported from middle management based on reluctance to learn about new technology or reluctance to face any upheavals in working methods.

Before the introduction of robots 13 per cent of the users say that the attitude of the workers directly affected was very favourable and a further 28 per cent quite favourable, compared with 9 per cent quite unfavourable and only 1 per cent very unfavourable, while a further two-fifths said the attitude of those expecting to be affected was neutral. Once the robots are actually introduced attitudes become more positive still: in 32 per cent of the plants the attitudes of those affected are said to have been very favourable and in a further 40 per cent quite favourable, while in only 4 per cent are they said to have been quite unfavourable or very unfavourable.

The plants planning large numbers of robots more often reported unfavourable attitudes in advance of their introduction than those planning only a single robot; however, after their introduction, the plants with larger numbers of robots report favourable attitudes even more often than those with only one robot.

Attitudes of workers affected

row percentages of robot users

	favourable	neutral	unfavourable	NA	TOTAL
Before robots	42	41	9	7	100
After robots	71	14	4	11	100

Four out of five robot users consulted the workers directly affected in advance of introducing robots, and in the plants which practised this consultation favourable attitudes were twice as common as in those which did not. Among these plants there was also a higher than average proportion with favourable attitudes after the introduction of robots. It seems probable that the advance consultation helped to enhance existing favourable attitudes and to counter unfavourable ones, although the figures could also be given the construction that the consultation was a *response* to potentially unfavourable attitudes rather than a *cause* of more favourable ones.

In about one quarter of the plants there have been specific negotiations with the trade unions about the introduction of robots. The trade union negotiations have been far more common in

41

Training, Employment and Industrial Relations

Effects on employment

(Tables 79, 81-87; Question 30)

Part of the explanation of the generally favourable shopfloor and trade union attitudes to robots lies in the fact that their introduction has not so far resulted in the devastating loss of jobs that some had expected. Three-fifths of the robot users in the sample report no change in the numbers employed by them as a direct result of the introduction of robots. One in four report a decrease in jobs, and one in 12 an increase, as a result of their use of robots.

The increases, although less common than the decreases, can be important. They occur mainly when the use of robots so improves a firm's competitive efficiency that sales are increased and extra workers are needed to produce the increased output. Usually the relationship is not easy to clarify and so is unlikely to have been reported by many of the firms in the sample. There are, however, some clear-cut examples of this, as in the case of the firm which was obliged to bring in robots as the only way to meet the volume and quality requirements needed to win a major new order. As a direct result an extra eight people were employed on the robot line itself and more than a hundred more elsewhere in the plant.

Effects of robots on employment

	increase in jobs	no change in jobs	decrease in jobs	TOTAL
Percentage of plants	8	60	25	7	100
Number of jobs	+ 700		– 1400		– 700

More common are the cases where the introduction of robots has led to some reduction in jobs — indeed, savings in labour costs are among the more important benefits which employers expect to get from the use of robots. These decreases have tended to be more common and larger in the bigger plants, in the plants with more robots, and in the plants which have been using robots over a longer period. However, the size of the decreases has tended to be smaller than often expected — typically about 8 jobs in each of the plants with a decrease.

When allowance is made also for the plants with no change in jobs as a result of their introduction of robots, and for those with an increase in jobs, the net overall effect is an average decrease of only about one job per plant in all the plants using robots, and a total net decrease of probably somewhere in the region of 700 jobs in all the robot-using plants in the country. And these figures reflect only the *direct* effects on employment as perceived by managers in the plants using robots. They do not take account of possible increases in employment by the robot suppliers or of the indirect economic effects of the improved productivity resulting from the use of robots.

The impact on shopfloor attitudes of the job decreases which have occurred has been much affected by the form which the decreases have taken. Actual redundancies due to the introduction of robots appear to have been very rare indeed. The normal practice, when the

Training, Employment and Industrial Relations

introduction of robots has led to job shedding, has been to make the change through natural wastage — waiting for people to leave of their own accord and then not replacing them — or through redeployment of the people not needed on the robot line to other work in the same plant. Hence there has usually been little reason for shopfloor opposition to protect job security, and shopfloor attitudes to robots are reported to have been favourable even in the great majority of the plants where there has been a fall in employment as a result of the introduction of robots. Where there have been any differences they have more often been to do with the conditions of the work itself or the remuneration for it.

It should be remembered, however, that the relatively small effects on employment *so far* are a reflection not only of the limited direct impact of robots — about one job shed for every four or five robots — but also of the comparatively small number of robots so far in use. When the deployment of robots and other forms of advanced manufacturing technology gathers momentum, as it needs to do, it must be expected that the impact on jobs will be correspondingly greater.

Training

(Tables 88-93; Questions 31, 32)

The use of robots requires specific training for the robot supervisors, and one way of ensuring that the introduction of robots does not lead to redundancies is by training the workers whose previous jobs have been displaced by the robots. In general, there does not seem to have been a problem in finding suitable people to train or in persuading them to take the training, although there do seem to have been some cases where employees have been concerned about possible de-skilling or reduced earnings. One of the production managers interviewed pointed out that some of the welders displaced by the firm's welding robots are now doing more monotonous — if safer — work in minding and feeding the robots. Some are unhappy at being 'machine-paced' and no longer on premium daily rates. The gain is 'that they are more or less guaranteed a job', and they are moved around the factory to relieve the monotony. But the majority of the interviews have revealed few problems in redeploying staff: for many 'displaced' workers there has been a shift to new skills in supervising, programming and operating robot systems, or they have simply been able to carry on doing their old job alongside the robots.

Three-quarters of the robot users have sent people on training courses provided by the robot suppliers and these account for about half of all the training course places used. About one quarter of the users put people through courses provided in-house, and these account for about one quarter of all the training course places. A few users make use of specialist training organisations, training facilities provided by other companies in the same group and other organisations providing courses. Altogether it seems that somewhere in the region of 7000 people have been on special robot training courses, an average of about ten people per plant with robots and about two people per robot in use. The number per plant is obviously greater in the plants with more robots.

There are three robot users satisfied with the training arrangements available for every one not satisfied, and it would seem that the heavy reliance on the robot suppliers to provide the training usually ensures that the training is relevant. Among the one in four plants not satisfied there appear to be few with serious problems.

Robot training courses

	own staff in-house	other company in group	robot supplier	specialist training organisation	other	TOTAL
Percentage of plants using courses	27	5	76	10	2	
Number trained on courses	1,800	500	3,500	600	700	7,100

9 DIFFICULTIES

Difficulties expected in advance
(Tables 94-99, 106; Questions Q.24, NQ.15)

The difficulty most commonly expected in advance was with the installation of the robot and its integration into the rest of the production process. The apprehension was largely justified in that problems with installation have indeed been common, although the proportion of users actually experiencing them (32 per cent) has only been three-quarters of the proportion expecting them (43 per cent).

The next most feared difficulty, expected in advance by 31 per cent of users, was opposition from the shopfloor and the trade unions. This apprehension has almost invariably proved to be unfounded, with only 2 per cent of the robot users actually experiencing difficulties on this account. The reasons for this are explained in the previous chapter.

Principal difficulties expected by robot users before they introduced robots

	percentage of robot users
Installation, integration	43
Opposition from shopfloor workers, unions	31
High costs of development	26
High costs of equipment	26
Lack of technical expertise	25
Reliability, maintenance	21

The third most common worry was over the levels of equipment costs and development costs (26 per cent each). In the event, problems with equipment costs have been slightly less common than expected (23 per cent of users), but high development costs have proved to be the most widespread problem of all, affecting more than one third of all the users of robots. These costs are often much higher than expected, and sometimes the purchase price is dwarfed by comparison. Some interviewees comment on the damaging effect of high development costs on future investment: senior management becomes highly sceptical, and in some cases robots fall into disuse once a particular task is superseded because the cost of developing a new application cannot be justified.

Lack of technical expertise was expected to present difficulties by one quarter of the users and problems with reliability and maintenance by one fifth of them. In both cases the apprehensions have been justified by events. Finally, smaller numbers of users foresaw a number of other sources of difficulty, such as inadequate after-sales support, unsuitability of robots for the tasks envisaged, no advantages over hard automation, no advantages over existing equipment, health and safety problems, and opposition from top management or other groups in the company. Most of these problems have in the event been encountered only by very small proportions of users. The one striking exception is inadequate after-sales support, which was perceived in advance as a problem by only 8 per cent of users but has turned out to be the second most common difficulty of all, affecting no less than 33 per cent of all users.

The expectations of the non-users, particularly those expecting to start using robots in the next two years, are broadly similar to those expected earlier by the current users. If the actual experience of those expecting to start within two years follows a pattern similar to that of existing users, there will be difficulties for more than the 19 per cent already expecting inadequate after-sales back-up and the 15 per cent expecting trouble over reliability and maintenance. On the other hand, many of the 63 per cent worried about high costs of equipment, and of the 24 per cent concerned about the robot's suitability for its task, will be pleased to find their apprehensions unjustified. Most of the 26 per cent expecting opposition from shopfloor workers and trade unions should find that their fears are unfounded.

Principal difficulties expected by potential users

	percentage of non-users with plans to use robots in next two years
High costs of equipment	63
High costs of development	30
Installation, integration	30
Opposition from shopfloor, trade unions	26
Lack of technical expertise	26
Unsuitability for task	24

Difficulties actually experienced

(Tables 94, 100-105; Question 24)

The difficulties and disadvantages actually experienced by robot users are varied, widespread and important. However, they need to be seen in perspective. Even the most common kind of difficulty is experienced by only a minority of users; one in four of the users report no difficulties at all; and, by and large, the robot users do not regard the difficulties as sufficient to cancel out the many benefits.

The three most commonly experienced kinds of difficulty, each affecting about one in three of all robot users, are the high costs involved in development work, inadequate after-sales support from suppliers, and problems with installing the robots and integrating them into the rest of the production system. The three next most common kinds of difficulty, each affecting about one quarter of all robot users, are lack of specialist technical expertise, problems concerning reliability and maintenance, and the high costs of the robots themselves and of associated equipment.

A further one in eight of the users have found robots unsuitable for the task intended for them, while one in eleven have found them to have no advantage over hard automation. There have also been a number of other difficulties, but none of them have affected more than 6 per cent of the users.

Principal difficulties actually experienced by robot users

	percentage of non-users with plans to use robots in next two years
High costs of development	37
Inadequate after-sales support	33
Installation, integration	32
Lack of technical expertise	27
Reliability, maintenance	26
High costs of equipment	23

Costs

(Tables 108-110; Question 18)

The high costs of robots and associated equipment have been a problem for 23 per cent of robot users, roughly the proportion expecting this, and the high costs of development work on applications has affected 37 per cent of users, more than any other kind of difficulty, and half as many again as had been expecting it. The high cost of robots is well known and an important barrier to wider use. Of the robot users in the sample, less than one third report that their latest robot cost under £30,000, and far more than one third of them report that their latest robot cost more than £50,000. The median cost was £39,000.

Difficulties

Among the users and potential users interviewed, the great majority say that lower costs would be the most helpful development in allowing them to make wider use of robots. Many express scepticism about the cost of robot systems ('There's a lot of padding in the prices'), and several say that robots cannot compete with cheaper pick-and-place devices in terms of cost-effectiveness for tasks such as machine loading.

Costs vary greatly with different kinds of application, and the simpler robots used in injection moulding mean that equipment costs tend to be below average in the plastic products industry, where one third of the machines cost less than £20,000 each. The heavier and more complex machines used in welding are largely responsible for the above average costs in the vehicles and metal goods industries, where about half the machines cost more than £50,000 each.

An important consideration is that, whatever the cost of the machine itself, additional outlays are required for the various kinds of associated equipment (grippers, attachments, guarding, feeders, storage, etc.) which need to be used with it, and also for the development work on applications. The sums involved are usually substantial and the problems they give rise to can be exacerbated by the fact that they are often inadequately allowed for in advance. Difficulties with development costs tend to be above the average in the larger plants, in mechanical, electrical and electronic engineering, and also in applications following a feasibility study carried out by a consultant.

An illustration of the costs involved in development of a robot *system* comes from one of the interviewed non-users, for whom casting is a potential application because of the hazards of the work. In order to replace a worker in the casting shop a robot would have to be bought, installed, programmed and tooled, the casting shop would have to be reconfigured, and the current processes and equipment would have to be redesigned to integrate them with the robot. For instance, a means would have to be found of automating the process of getting hot metal to a cast and of measuring the correct volume. This case shows how a robot must be thought of as part of an automated system: the cost of the robot alone would be outstripped by the costs of developing it as a system, but the robot on its own would be no solution.

Installation, expertise and after-sales support

Inadequate after-sales support has been a problem for 33 per cent of robot users, the second most common problem encountered, and the more serious in that for three out of four of the plants concerned it is a problem which had not been expected. It is more common in the plants with the larger robot installations and it is probable that more than 40 per cent of all the robots in the country are in plants where this is considered a problem. Interviewees complaining about after-sales service often express the view that some suppliers do not have enough skilled maintenance staff to cover their full range or to do justice to the needs of all clients.

The importance of after-sales support lies largely in the fact than many users depend on their suppliers for expert technical back-up because they have problems with installation and reliability and because they do not have the required level of technical expertise in-house. The series of PSI national industry surveys (1981, 1983 and 1985) has found lack of technical expertise to be the most common difficulty encountered in the use of microelectronics technology in general. In robot applications it is clearly also a matter of importance with the 27 per cent of all users citing it as a difficulty. Both installation problems and difficulties through lack of technical expertise tend to be experienced with above average frequency in the plants with the larger robot installations, in the vehicles industry, and in the plants which used a consultant for their feasibility study.

Lack of technical expertise and poor planning often combine to create serious installation problems. The introduction of a robot demands planning changes to the environment in which it will work: plant layout, jigging, components, upstream and downstream processes. One user points out that robots can make things *worse* for a firm 'if they are introduced into the same old shambles as before' and if components and fixtures are not brought up to the required levels of quality and consistency. Numerous interviews underline the fact that poor planning of the use of a robot as part of a manufacturing *system* can lead to high development costs and difficulties with installation.

Difficulties

Reliability and maintenance (Tables 94, 100-106, 111-120; Questions 21, 22, 24)

Problems with reliability and maintenance, experienced by 26 per cent of robot users, take on increased importance in the absence of strong in-house technical expertise or adequate after-sales support by suppliers. These problems tend to be more common in the bigger installations, in the larger plants and in the vehicles industry.

One measure of the extent of this kind of problem is the amount of downtime experienced. Altogether 37 per cent of the robot users have had more downtime than expected, compared with only 23 per cent which have had less than expected. Greater than expected downtime appears to be particularly common in the mechanical, electrical and electronics engineering industries, in the smaller robot installations, particularly those with only a single robot, and in the plants where the feasibility study was done by a consultant.

The causes of downtime are varied, the most common being not the robot itself but the associated equipment (a frequent or very frequent cause for 35 per cent of users). The other main causes are the robot itself (24 per cent), unsatisfactory components or materials (20 per cent), and other equipment upstream or downstream (17 per cent). Half the users say they only occasionally have downtime caused by trouble with the robot itself. About one third say they only occasionally have downtime through each of the other causes, but hardly any say they never have any unintended downtime at all. Several interviewees said that downtime had been much more than expected at the outset, but reported that the problems had diminished the longer the robot systems had been running. In some cases planning had been poor: after installation of the robot unsuitable components or materials led to lengthy delays as parts had to be redesigned to allow effective robot operation.

10 BENEFITS

Main benefits experienced
(Tables 121-128; Questions Q.23, NQ.14)

In general, the benefits achieved from the use of robots have been very similar, in terms of rank order, to those expected in advance, but the proportion of users reporting each has tended to be only about three-quarters of the proportion expecting it. Nevertheless the proportions experiencing the main benefits have been nearly double the proportions experiencing the main difficulties. Each of the five most common benefits has been experienced by about half of all the users and the six next most common ones have each been experienced by between one quarter and one seventh of all the users.

Principal benefits experienced by robot users

	percentage of robot users
Improved quality, more consistent products	58
Lower labour costs	52
Greater volume of output	44
Improved work conditions, environment, safety	44
Increased technical expertise	43
Better management control	23
Greater reliability, less downtime	18
Better labour relations	17
Greater flexibility for product changes	16
Lower material costs, less waste	16
Less capital tied up in work in progress	15

While typically the users report three or four benefits each from their use of robots, one in five say they have experienced no benefits at all. However, these are mainly accounted for by the one in five of the users whose robots are still in the experimental or pre-production stage. Of those whose first robot is already installed for commercial production, only 7 per cent report no benefit experienced.

Effects of plant size
(Table 123; Question 23)

There is no uniform general pattern in how the benefits of using robots vary with the size of plant, but there are clear size-related differences in some of them. For example, the smaller plants have benefited disproportionately from improved quality and more consistent product, from greater volume of output, and from greater reliability and less downtime. The larger firms have benefited disproportionately from better managerial control, less capital tied up in work in progress, lower material costs, less waste and improved work conditions and safety. Broadly speaking, there seems to have been some tendency for the smaller plants to get more technical benefits and for the larger ones to get the more 'managerial' ones.

Effects of industry
(Table 124, Question 23)

The pattern of benefits achieved in different industries is also not uniform, but in general it seems that users in vehicles and other metal goods have tended to get more benefits than robot users in the other main industry groups, particularly in the form of improved quality, lower labour costs, improved work conditions, increased technical expertise, and greater flexibility for product changes.

Effects of length of use
(Tables 125, 126; Question 23)

There is a clear tendency for the benefits of robot use to increase with time. For example, those who started to use robots only in 1985, many of whom had not yet got their first robot into production at the time of the survey, report experience of benefits markedly below the average, those who started in 1984 about the average, those who started in 1981-1983 well above the average, and those who started before 1981 far more again.

Benefits after different periods of use of robots
percentage of robot users getting benefit

	first robot in 1984	first robot before 1981
Improved quality, more consistent product	64	85
Lower labour costs	50	73
Greater volume of output	37	70
Improved work conditions, environment, safety	46	76
Increased technical expertise	49	61
Better management control	20	42
Greater reliability, less downtime	11	27
Better labour relations	11	27

Number of shifts
(Table 127; Question 23)

It is also striking how almost all the benefits are more common in plants with more than one shift. This applies particularly to the benefits related to labour costs and output.

Benefits with different numbers of shifts
percentage of robot users getting benefit

	one-shift plants	multi-shift plants
Improved quality, more consistent product	57	60
Lower labour costs	36	61
Greater volume of output	37	48
Improved work conditions, environment, safety	38	48
Increased technical expertise	33	49
Better management control	15	27

Effects of number of robots
(Table 128; Question 23)

Perhaps the most significant aspect of the pattern of benefits is the way that most of them are experienced more often in the plants with 2–5 robots than in those with only one, and more often still in those with six robots or more.

Benefits experienced with different numbers of robots
percentage of robot users getting benefit

	robots in use:		
	1 robot	2-5 robots	6+ robots
Improved quality, more consistent product	48	61	74
Lower labour costs	43	54	72
Greater volume of output	37	44	63
Improved work conditions, environment, safety	33	52	58
Increased technical expertise	33	43	70
Better management control	15	24	42
Greater reliability, less downtime	14	15	30
Better labour relations	17	12	30
Greater flexibility for product changes	12	19	23
Lower material costs, less waste	6	20	30
Less capital tied up in work in progress	7	13	33

Benefits

Profitability

(Table 129; Question 26)

Ultimately, what matters is not how many benefits the robot users enjoy, or how many difficulties they endure, but whether, overall, the use of robots has actually made the enterprise's operations more profitable. On this test the answer is clear — a large majority of all robot users have found that their robots have contributed to increased profitability.

Moreover, figures for all users include experimental machines and newly delivered ones not yet on-line for commercial production, and not expected at this stage to make a contribution to the company's profits. If the calculation is limited to plants with robots already installed for production, the users who have found their robots profitable outnumber those who have not by more than three to one.

Effect of robots on profitability

	percentage of robot users	
	all users	users with robots installed for production
Has robot made operations more profitable?		
Yes	61	70
No	24	23

Intention to buy more robots

(Table 135; Question 33)

In some ways the acid test of whether the benefits arising from the use of robots are sufficient for them to be considered successful is whether the firms which are already using robots intend to get any more of them in the future. On this test the margin of success is clear: of the robot users in the sample, 60 per cent say they plan to buy more robots within the next two years, and only 33 per cent say they do not.

It should be stressed that this two-to-one majority is likely, if anything, to understate the degree of success. If the use of robots has proved to be a mistake a firm is unlikely to compound the failure by buying more of them. On the other hand, if the use of robots has been a success, there could still be reasons why, despite this, a firm does not plan to buy more of them within the next two years. It may, for example, already have enough robots to meet its needs; or it may wish to give priority to investment in other areas; or it may for the moment lack the financial resources to make further capital investments of any kind.

Overall value

(Table 129, Question 25)

It may therefore be felt that intention to buy more robots is not a fully satisfactory measure. And it can be objected that not even profitability is an adequate test, since some users may not be in a position to measure the profitability of a single, possibly new, machine. Moreover, some may have used their robots in ways which have proved useful but which had never been intended to make a contribution to the company's profits at this stage. If, therefore, the more general question is asked, whether overall the use of robots has proved worthwhile, the answer is overwhelming — for every user who considers that their robots have not been worthwhile there are no less than 28 who consider that they have been worthwhile.

Whether use of robots has been worthwhile

	percentage of robot users
Very worthwhile	49
Fairly worthwhile	32
Marginal	10
Not really worthwhile	2
Not at all worthwhile	1

11 FACTORS IN SUCCESS

If the general experience of using robots has been successful, the fact remains that some users have enjoyed much more success than others. It is therefore of interest to review the findings of the survey in order to identify which factors appear to be associated with above the average degrees of success. Here it needs to be stressed, first, that many of the factors are interrelated, and too much attention should not be paid to any one of them in isolation; secondly, like most generalisations, they do not apply in all cases — however great the majority in one direction, the minority in the other direction is no less real for being small.

Nevertheless, an examination of the factors associated with greater than average success does provide some indication of the probability of success attending different characteristics and courses of action. This may accordingly be of value to potential users wondering whether the use of robots is likely to be successful for *them*.

The survey provides three measures which can be used as indicators of success: whether the use of robots is considered to have been 'very worthwhile', whether the use of robots has made the firm's operations more profitable, and whether the firm plans to buy more robots in the next two years. The tables in the following sections give, for each of the factors considered, the percentage of existing users who have found robots very worthwhile, or profitable, or who plan to get more of them. These percentages are given in **bold type** when they are 5 percentage points or more above the percentage for the sample of users as a whole, respectively 49 per cent finding robots very worthwhile, 61 per cent finding them profitable and 61 per cent planning to get more of them.

Plant characteristics

Table 130; Questions 25, 26, 33)

Factors in success

row percentages of robot users

Plant characteristic	robot very worthwhile	robot increased profits	plan to get more robots
Employment size			
1–99	35	48	48
100–999	53	64	59
1000 +	47	60	**68**
Industry			
Mechanical engineering	40	54	47
Electrical, electronic instrument engineering	53	56	64
Vehicles, aircraft	**55**	**69**	**69**
Ownership			
UK	47	59	61
Overseas	**57**	**68**	62
Number of shifts			
One	36	46	47
Two	**54**	**67**	63
Three	**58**	**73**	**78**
ALL ROBOT USERS	49	61	61

Factors in Success

The small plants, employing less than 100 people, score well below the average on all three measures, probably a reflection mainly of their tendency to have fewer areas with scope for applications and less technical expertise with which to exploit them. The medium size plants, employing between 100 and 1,000 people, have found robots very worthwhile and profitable marginally more often than the large plants employing more than 1,000 people. The large plants, however, include a proportion well above the average planning to get more robots. Here there are a number of mutually reinforcing influences at work: the larger plants tend to be the ones which started earliest and have the greatest numbers of robots already, both of these factors being associated with above average intentions of buying more robots in the next two years.

There are percentages well above the average on all three measures in the vehicles industry, associated with early use, large installation, good scope for applications and the expertise to implement them successfully. More ominously, there are also percentages well above average for worthwhileness and profitability in the plants in companies which are part of overseas owned groups.

The plants which work only a single shift in the sections using robots score well below the average on all three measures and the plants which work two or three shifts score well above the average. These differences are important because they are larger than the others but also because, unlike the other characteristics, they are a feature of the plant which is potentially capable of being changed. Prospective robot users who at present work only a single shift would seem well advised to investigate whether there is scope for improving the chances of success by going over to a two shift system in the section in which the robots are to be used.

Introduction of robots (Table 131; Questions 25, 26, 33)

The companies which devolve to plant management the decision to buy the first robots do not appear to find them any more profitable subsequently, although they may be more inclined to buy more of them.

Factors in success

row percentages of users

Features of introduction	robot very worthwhile	robot increased profits	plan to get more robots
Point of decision			
Company board	48	61	59
Plant management	53	60	**72**
Feasibility study			
In-house	52	58	**67**
Company in group	25	**67**	**69**
Robot supplier	**58**	**67**	65
Consultant	52	49	45
None	46	63	38
Installation			
Turnkey by supplier	52	65	57
Naked robot	49	58	**66**
Price of first robot			
Under £20,000	52	**74**	**76**
Over £50,000	49	59	37
Workers' attitudes (after introduction)			
Favourable	**60**	**66**	**66**
Unfavourable	11	22	33
Year first robot acquired			
1980 and before	**76**	**85**	**76**
1983 and after	41	54	58
ALL ROBOT USERS	49	61	61

The approach adopted to feasibility studies and installation appears to have a greater bearing on subsequent success. Those having a feasibility study undertaken by the robot supplier have fared rather better than the average, while those using a consultant score markedly below the average on all three measures. Nearly two-thirds of those not undertaking any feasibility study at all have nevertheless found robots profitable and nearly half have found them profitable (in each respect, about the average for all users), but not much more than one third of them have plans to buy more robots (well below the national average).

Those adopting a turnkey approach to installation have been marginally more successful than the average in terms of worthwhileness and profitability, but those buying their first robot on its own are marginally more inclined to buy more robots in the course of the next two years.

The fact that the firms with no feasibility study and the firms buying the robot on its own have not done all that much worse than the others should not be taken by potential users to indicate that feasibility studies and turnkey packages will make little difference. It must be remembered that some of the firms which undertook no feasibility study and bought their first robot on its own did so not from weakness and inexperience, but from strength and confidence: it was because they reckoned they had the in-house expertise and experience to feel sure they knew what they were doing that they felt able to go it alone. Normally, however, any company planning to use robots for the first time, unless it has strong, relevant expertise, is likely to face much less risk of failure if it arranges for a proper feasibility study and a turnkey approach to installation.

The firms with the less expensive robots have more often found them profitable and decided to buy more than those with the more expensive ones. This is probably not so much an indication that those who managed to pay less have done better out of it, but rather that the kinds of application, such as injection moulding, which can be performed by the less expensive kinds of machines have tended to be more economically successful.

Unfavourable attitudes of the workers affected *before* the introduction of the robots does not normally seem to have been prejudicial to the outcome subsequently, but in the very rare cases where attitudes have still been unfavourable *after* the introduction of the robots, the investment's lack of success has been dramatic. Half of these plants have found their use of robots not worthwhile or marginal, three-quarters of them have not found robots profitable and two-thirds of them do not plan to get any more of them. While it must be stressed that only very few robot users have found themselves in this situation, it is clearly relevant to note that the practice of advance consultation with those affected has been associated with higher proportions of favourable attitudes.

It is striking that the proportion of users who have found their use of robots profitable rises year by year from 47 per cent of those who started only in 1985 to 85 per cent of those who started in 1980 or before. The practical message of these figures is not that there is anything special about any particular start-up year but rather that the longer a firm uses robots the more profitable they are likely to be. In the first year earnings are likely to be limited because of the time needed to install the machine and develop the first application. Thereafter, year by year, as experience increases, the robot is likely to be used with increasing efficiency and increasing benefit. Hence the two or three year pay-back periods commonly insisted on by company financial controllers would seem particularly inappropriate, since for first installations or new applications it is usually after longer periods that the main economic benefits accrue. Robots subsequently introduced for the same or similar tasks are likely to be installed and moved into production faster and more easily. For these a two- or three-year payback may well be a more realistic proposition.

Factors in Success

Number and application of robots

(Table 132; Questions 25, 26, 33)

Factors in success

row percentages of users

	robot very worthwhile	robot increased profits	plan to get more robots
Number of robots now			
1	37	47	29
2	48	64	50
3–5	**60**	**68**	**74**
6+	**72**	**86**	**77**
Area of application (plants using robots on application specified)			
Arc welding	47	58	51
Spot welding	**62**	**69**	**62**
Painting, coating	**58**	**78**	**72**
Glueing, sealing	**59**	59	**71**
Assembly	46	50	**69**
Handling	53	**71**	**74**
Grinding	50	**68**	**79**
Injection moulding	**55**	**70**	**85**
Press loading	39	**77**	**77**
Machine loading	51	**78**	62

The degree of success rises steadily with the number of robots used just as it does with the length of time over which they are used — indeed the two tend to go together. The plants with six robots or more have found their use of robots very worthwhile and profitable twice as often as those with only a single robot and more than twice as many of them are planning to buy more. The plants with the larger installations tend not only to be able to support more specialist maintenance staff but also to gain important economies through being able to hold stocks of more key spare parts, or even complete reserve robots. These help reduce the frequency and duration of breakdowns and the costly production hold-ups resulting from them.

It does not follow from this that every potential user should think in terms of starting off with at least six robots. But in so far as the larger installations are so much more often successful, it would seem sensible for new users to think in terms of applications which at least have the *potential* for multiple installations in due course, even if they may sometimes prefer to build up experience with just a single machine initially.

All the most common areas of robot application (except arc welding) show above average scores on at least one of the three measures, and several are markedly above average in all three — a reflection of the fact that they have become the most common areas of application largely *because* they have been found to be successful. New entrants to these areas can therefore count on abundant successful experience to build on. It follows, of course, that in many of the other areas of application, at present used by only small numbers of users, the success rates are well *below* average. This is partly because many of these other areas of application have been exploited only recently and have not been used for long enough for all the teething problems to be overcome and for all the benefits to show; and in some cases it is because the applications are inherently less suited to robots, which may indeed be the reason they have not been used earlier. It is likely that, as time goes on and experience builds up, many of the newer and less common applications will become as generally successful as the more strongly established ones are already. Meanwhile, however, possible new users of robots will need to be more cautious in going into these areas of application unless they are confident of having the requisite expertise for success.

Benefits and difficulties

(Table 133; Questions 25, 26, 33)

As might be expected, there is a high correlation between the robot users who experienced the main benefits identified and those which were successful in terms of the three measures used; and, conversely, between those who experienced the main difficulties identified, and those which were less successful.

Factors in Success

Improved quality of products, lower labour costs, greater output, better work conditions, greater control, increased expertise and more reliability are all associated with above average proportions of users finding robots very worthwhile and profitable and planning to get more, with some of them strikingly above average in all three respects. Hence any prospective user with grounds for expecting major benefits in these areas may take encouragement from the knowledge that they tend to be associated with success in a more general sense.

Likewise, the plants which have had difficulties with installation, reliability, downtime, equipment costs, development costs, and after-sales support have been less successful than the average in terms of the measures used as indicators of success. It must be stressed, however, that what is being measured, for the most part, is not success against failure, but relative degrees of success. Even of these relatively less successful robot users, over half have found their robots profitable, more than three-quarters have found them fairly worthwhile or very worthwhile, and more than half plan to buy more robots within the next two years.

12 PLANS AND PROSPECTS

Existing robot users
(Tables 134–142)

The existing robot users in the survey have given information about their future plans, and two out of three say they expect to acquire more robots in the course of the next two years. Altogether their plans suggest an increase of about 60 per cent in the total number of robots in the course of only two years. There are a number of grounds for doubt as to whether this will in fact be achieved, but even if the size of the increase turns out to be smaller than expected, it seems likely that the *pattern* of the increase may well be similar to that suggested by the survey respondents.

In all industries there are more existing users expecting to acquire additional robots than there are users which are not, but in the vehicles industry the former exceed the latter by more than two to one, and the average number per plant is also above the average, with the result that the industry accounts for more than one third of all the additional robots expected by existing users in the next two years. In the plastics industry the intention to expand is even more widespread, with those expecting to acquire more robots outnumbering those who are not by four to one, but the average number of additional robots per plant is less than in the vehicles industry and, because of the much smaller base, the total number of additional robots expected is much less than in vehicles.

The overseas-owned plants differ from the British owned ones in that a higher proportion of them expect to acquire more robots, in substantially greater numbers, with the result that the average increase in robot numbers per plant is expected to be two-thirds greater than in the British owned plants. Altogether the overseas-owned plants are expected to account for more than one quarter of the total additional robots acquired by existing users.

While more robots are expected by many of the plants in all size ranges, the ratio of those which expect to acquire more robots to those which do not rises from 50 : 48 in the smallest plants employing less than 100 people to 66 : 27 in the largest ones employing more than 1,000 people. These large plants also expect to acquire four times as many additional robots per plant as the small ones, with the result that they are expected to account for about 60 per cent of all the additional robots acquired by existing users.

In the plants using robots only on a single shift, the numbers expecting to acquire more robots are similar to the numbers not expecting to; in the plants using robots on two shifts the former outnumber the latter by two to one; and in the plants using them on three shifts they outnumber them by four to one. The plants using robots on more than one shift also expect to acquire more robots in bigger numbers than those using them on single shifts, with the result that they are expected to account for 80 per cent of all the additional robots acquired by existing users.

The plants which have been using robots over a longer period are more inclined to plan to acquire more of them in the future, and in substantially greater numbers, than those starting more recently. For example, the plants which started before 1981 expect to acquire on average four times as many more robots each as those which started only in 1985.

Plans and Prospects

The plants with greater numbers of robots already are also more likely than the others to expect to acquire more robots in the next two years, and in substantially greater numbers. Thus the plants which already have six or more robots expect on average to acquire eight times as many more each as those with only one robot already. Those with more than one robot already are expected to account for more than 80 per cent of the total additional number acquired by existing users.

Not surprisingly, the users who have found their use of robots very worthwhile are much more inclined to acquire more of them than those who have found them only marginally, or not, worthwhile (79 per cent against 29 per cent); and those who have found their robots profitable already are more inclined to buy more of them than those who have not (74 per cent against 37 per cent). The former groups also intend to acquire more robots in greater numbers, with the result that they account for about 60 per cent and 75 per cent respectively of the total increase in robot numbers expected by the existing robot users.

Numerous users have plans to introduce *related technology* such as pick-and-place machines or automated guided vehicles (AGVs) rather than buy more robots. Users who plant to buy more pick-and-place machines say that robots cannot compete with them for cost-effectiveness in applications such as machine loading. Others, while not dissatisfied with robots, see more possibilities for using flexible manufacturing systems (FMS) in which robots will not necessarily have a role.

Prospective robot users
(Tables 143–150)

The sample of non-users in the survey may be regarded as for the most part potential users of robots in that they are manufacturing companies which sent representatives to the Automan exhibition where they expressed an interest in industrial robots. However, if the undertaking of a feasibility study is regarded as an indicator of the possible imminence of actually acquiring robots, then the majority appear still to be some way off. Altogether 63 per cent say that no feasibility study has yet been undertaken and none is planned, 11 per cent plan to do a study in the future, a further 4 per cent have one currently in progress, and 21 per cent have one already completed. Of these latter, however, only 9 per cent recommended the use of robots compared with 12 per cent which did not.

The percentage of plants with a feasibility study already completed is twice as great in plants working more than one shift as in the single shift ones, and three times as great in the larger ones employing over 500 people as in the smallest ones employing less than 100 people. It is also above the average in the vehicles and metal goods industries. Only in vehicles, however, have the studies shown a clear majority in favour of using robots. Half the plants which have had a feasibility study which recommended the use of robots, and half of those which have a study planned, say they intend to acquire one or more robots in the next two years; and the other half say they have the use of robots 'under consideration'. Hardly any of the others definitely plan to use robots in the course of the next two years, although half the total sample say they have the use of robots 'under consideration'.

Altogether 15 per cent of the plants in the non-user sample say they expect to acquire robots in the next two years — on average about three each — and this may be regarded as indicating a total acquisition of somewhere in the region of 2,000 robots by new users over a two-year period. However, as with the additional robots expected by the existing users, the answers of the survey respondents are probably more reliable as an indicator of the *pattern* of future robot use than of its *size*.

In general, the pattern of prospective new robot users is similar to that of existing users; higher proportions of plants planning to acquire greater numbers of robots in the larger plants, in the multi-plant companies, and in the plants working more than a single shift. The percentage of plants expecting to acquire robots is above the average in the vehicles and metal goods industries, and the number of robots planned per plant is far above the average in the plastics industry.

Growth prospects
(Tables 151, 152)

The answers of the survey respondents may be used to get some indication of the growth in the total number of robot users to be expected in the future. If it is assumed that all those who

Plans and Prospects

say they expect to acquire robots for the first time in the next two years will in fact do so (but that none of the other non-users will do so), and if allowance is made for the other visitors to the Automan exhibition who did not reply to the questionnaire or were not included in the sample, and if it is further assumed that only about half of all potential robot users sent representatives to Automan, then it can be calculated that somewhere in the region of 800 additional plants will become robot users in the course of the next two years, more than doubling the total number.

However, it can be argued that these assumptions lead to a figure which is much too high. It may be that there are no substantial numbers of imminent users in addition to those who came to Automan, and it may be that many of those who say they will use robots within the next two years will in the event not do so. In the 1985 PSI survey of microelectronics in industry many establishments said they expected to be using robots in two years' time, but when re-approached one year later only 20 per cent of them were in fact already doing so. Also, robot suppliers for the most part report that sales during 1986 have been less than hoped. If it is assumed that *all* those likely to start using robots in the next two years went to Automan, and if it is assumed that only 40 per cent of those saying they will start using robots within two years in fact do so, this provides the basis for alternative low assumptions giving the very much smaller figure of only about 160 additional users, an increase of about one fifth over two years.

Change in number of plants using robots

Assumptions	existing users	new users	total in two years
Low	740	160	900
High	740	800	1540

For particulars of high and low assumptions see Appendix I.

Similar estimates, on the basis of alternative high and low assumptions, can also be made for the total number of robots that will be in use two years after the time of the survey. If the expectations of survey respondents are taken at face value (and allowance made also for those not included in the survey), it can be calculated that existing users and new users will each acquire about 2,000 robots, bringing the total number in use to more than 7,000, an increase of more than 120 per cent over two years. On the other hand, on the alternative low assumptions that *all* prospective users were at Automan, that only 40 per cent of those expecting to start using robots will in fact do so, and that only one third of the plans of existing users for more robots will actually be realised, an addition of only somewhere in the order of 1,000 robots may be expected, an increase of about one third over two years.

Change in number of robots

Assumptions	existing users	additional existing users	additional new users	total in two years
Low	3200	6509	400	4250
High	3200	1950	2100	7250

For particulars of high and low assumptions see Appendix I.

It is clear that there is a very wide spread between the figures resulting from these alternative sets of assumptions. Yet it should be stressed that neither set of figures is fancifully unrealistic. The low figures are consistent with the disappointed expectations of the establishments in the earlier PSI survey; with the reduced rate of increase in robot numbers experienced in 1985; and with the poor robot sales reported by suppliers in 1986. The higher figures, on the other hand, represent no more than what the firms themselves say; are no more than the percentage rate of increase already achieved in the two-year periods 1981–1983, 1982–1984 and 1983–1985; and constitute an increase in absolute numbers similar to that already achieved in West Germany between 1983 and 1985.

Whether the rate of growth will turn out to be nearer the lower figures, which at present seems more probable, or nearer the higher figures, which should potentially be attainable, will depend on a number of factors; in particular, on the policies adopted by the current and prospective users of robots, by the robot suppliers and by the government. Some indication of which policies will be relevant is provided by the assessment of the robot users themselves of what will most help them make effective use of robots in the future.

Requirements of robot users

(Tables 153–155; Questions 34, 35)

In the view of the firms already using robots, the most important obstacles to more effective use of robots are financial and economic ones. These account for four of the seven most widely sought improvements. No less than 70 per cent of existing users say that cheaper robots would be particularly helpful in enabling them to make better use of robots, and 53 per cent ask for cheaper associated equipment. 50 per cent of the users call for more government support, and 21 per cent for easier finance for investment (to help meet these capital costs) and 41 per cent for an upturn in the economy (to help provide the expanding markets to justify the expenditure). In addition, 21 per cent would like lower operating costs.

Although lack of capital is often the main obstacle to use, many prospective users are held back by problems of cost justification related to the potential applications. A plant may be fully satisfied with dedicated automation for a task which could be done by robots; volumes and batches may be considered too small to justify the investment in a robot and the need to reprogram and retool for different products; or the potential tasks are done better manually than a robot could do them (this applies especially to precision assembly or sorting of parts, for example). In such cases there is no technical obstacle to using robots: but they cannot do the job better or more cost-effectively.

The next most widespread group of requirements identified by existing users are changes to make robots easier to use and to circumvent the deficiencies in specialist expertise felt by many of them. 56 per cent of them seek easier programming, 43 per cent ask for less need for special skills, 39 per cent for easier maintenance, 35 per cent for greater reliability, and 19 per cent for easier installation. Linked to these needs are a wish for better service from suppliers, with 26 per cent calling for better after-sales service, 11 per cent for quicker delivery and 29 per cent for more UK-based manufacturers. There are also calls for a change in the marketing strategy of robot suppliers. A comment made by numerous interviewees and survey respondents is that suppliers should market robot *solutions* rather than robots 'in isolation'; they should be sold as part of *systems* for automating factory processes. Typical comments are as follows.

> 'Suppliers peddle their wares in isolation.'
> 'They're doing themselves a disservice by concentrating on selling a robot instead of a system.'

There is also some impatience with what is seen as 'hype' about robots as the automatic answer to manufacturing problems. Several users complain that suppliers often fail to consider clients' production circumstances sufficiently in their eagerness to make a sale; for instance, by pushing robots which in fact offer more facilities than the client's real needs demand.

The third group of requirements, as seen by robot users, is improvements in the performance of the robots themselves. 50 per cent want better sensors (improved robot vision facilities in particular), 37 per cent greater speed, 31 per cent more accuracy, 29 per cent increased versatility, 19 per cent heavier payload and 11 per cent smaller, lighter robots. Those who say they would like more sophisticated robots outnumber by four to one those who would like robots less sophisticated than the ones they have already, but about half the total expect their next robots to be of about the same degree of sophistication as their present ones.

It does not follow, of course, that all the users who want improved performance in their robots are aware of the costs of providing it, still less that they are willing and able to pay for it. On the face of it, what they feel to be necessary is better, easier-to-use equipment, with more effective back-up — but at lower cost. This presents a challenge to robot suppliers — and substantial market opportunities to those who can meet it successfully. It also highlights the relevance of the government support policies which (until their recent withdrawal) made a useful contribution towards bridging the gap between the cost of what the suppliers could provide and what the users could afford.

13 POLICY IMPLICATIONS

There are a number of policy implications in the findings of the study which apply specifically to the users of robots, current and prospective, to the suppliers of robots, and to the government. But first there are two general findings which apply to all three.

The first is that, on the evidence of the study, which is the experience of the firms actually *using* robots, the use of robots has been a success. A wide range of important benefits has been secured by high proportions of users, and by three separate measures the use of robots is regarded as advantageous. Altogether, 81 per cent of robot users regard their use of robots as having proved worthwhile, with 49 per cent finding them very worthwhile; this compares with only 2 per cent who see them as not really worthwhile and 1 per cent who see them as not at all worthwhile. 61 per cent say robots have made their operations more profitable compared with 24 per cent who say they have not (or at any rate not yet, since many of these are new ones not yet in commercial production); and 61 per cent plan to buy more robots within the next two years. This evidence of success should encourage prospective users to go ahead with confidence; encourage suppliers to provide for the market with confidence; and encourage government to ensure that the obvious potential is effectively exploited.

The second general finding is that, if robots have been successful, it is for the most part as a key element in wider manufacturing systems using advanced technology. In formulating their different policies, users, prospective users, suppliers and government need to consider robots in the context of the integration of manufacturing processes and look at how they fit into strategies for investment in advanced technology in general. While the study has focussed mainly on the robots themselves, largely because they are a conveniently identifiable unit for measurement, it must be remembered that most of the findings apply really not so much to the robots themselves as to the advanced manufacturing technology systems of which they are a part.

Robot users

There are a number of findings from the study which have implications for robot users and, more importantly, for the much greater number of firms considering whether to become robot users and, if so, how best to set about it.

Awareness Companies contemplating using robots need to be fully aware of the specific benefits which can be provided by them, and clear about which of them they expect to achieve, so that they can plan to exploit them effectively. No less they need also to be aware of the various difficulties they are liable to encounter so that as far as possible these can be foreseen and averted or provided for. In this they need to draw on the advice of suppliers, the experience of existing users and the help available from government, particularly that offered under the AMT programme. They also need to draw on the many other sources of information and support open to them, for example: the trade and the technical press, exhibitions, conference, courses, professional institutions, industry associations, the National Economic Development Office (NEDO), the British Robot Association (BRA), the Production Engineering Research Association (PERA), the Open University, the Open Tech and local universities, polytechnics and higher education colleges.

Feasibility Introduction of robots without a feasibility study is likely to be risky unless a firm has a high level of specific expertise: on average the firms undertaking no feasibility study have taken half as long again to get their first robot into commercial production. Feasibility studies undertaken in-house have often proved satisfactory in big plants with relevant expertise. Smaller plants have tended to use the robot suppliers to do the feasibility study and those following this course have subsequently found their use of robots worthwhile and profitable slightly more often than others. Some firms have used consultants, usually with below average results. Half of those using consultants rated their reports as only 'fair', or worse. It seems that consultants need to be selected with care to ensure they have specific knowledge of the application of robot systems to the production task which is planned.

Installation and development High equipment costs are a common problem, but one which is generally expected and allowed for. The high costs of developing an application are much less commonly foreseen and therefore can give rise to unexpected problems. Development costs often amount to more than twice the purchase price of the robot itself; development involves (for instance) robot gripper development, component transport systems, safety equipment and redesign of components. These costs are the most frequently experienced difficulty of all, experienced by 37 per cent of all first-time users.

Well-planned installation is crucial to success, since the robot needs to be efficiently integrated into the rest of the production system. Installation and integration difficulties are the third most common type of problem encountered and are experienced by one third of all users; and the robot cannot start earning until they are overcome.

To help overcome these difficulties many robot suppliers offer a complete turnkey package. It enables the prospective user to draw on the supplier's expertise and familiarity with the equipment to help ensure a well-planned installation. It is usually the most prudent approach for the first-time user, or for further installations if these are in different application areas. Some users go it alone, buying only the robot on its own. Some do this because they have the in-house expertise to plan the installation successfully; and some purchase a robot solely for experimental or research purposes, to explore various potential uses. Others do it just to save the additional cost of the turnkey package, but they usually pay more in the end through underestimating development costs and/or making mistakes.

Payback Installation and integration of the first robot takes on average a little over six months before it comes on-line and starts earning. Thereafter its efficiency of use improves with experience year by year, and its maximum profitability is usually not achieved until several years after introduction. Hence the two to three year payback period commonly demanded by financial controllers is inappropriate, at least for initial applications, since it ignores the costs of learning how to apply the technology and does not take indirect or qualitative benefits into account. A realistic appraisal, at least for initial robot applications, requires the longer time perspective and strategic view normally found in countries like West Germany and Japan. It also demands allowance not merely for savings in labour and other direct production costs, but also for indirect savings, for example through reduced waste and more consistently high quality product (the most frequently experienced benefit from robots) and for more intangible benefits found by users, such as improved work conditions, better labour relations and a more modern company 'image'. (Several interviewees have said that existing and potential customers were impressed that robot systems had been introduced.)

Number of robots Profitability tends to increase markedly with greater numbers of robots in a single installation: altogether 37 per cent of users with a single robot have found it 'very worthwhile', compared with 72 per cent of those with 6 or more robots; and 47 per cent of those with a single robot have found it made their operations more profitable, compared with 72 per cent of those with 6 or more robots. While it is sensible to start with a single robot to assess the potential and limitations in a selected application, a greater financial return is obtainable if the experience gained can be fully exploited by investment in a number of robots to work on the same application.

Number of shifts A number of the benefits of using robots, particularly those associated with lower labour costs, are more frequently achieved in plants which work more than a single shift. Indeed, the ability of robots to work round the clock is one of their major potential strengths. Overall, the plants using robots on two or more shifts are much more often profitable and much more often considered the use of robots 'very worthwhile' than those working only a single shift. Plants which operate on a single shift basis need to consider seriously the possibilities for the integration of one-shift tasks into a two-shift robot application.

Policy Implications

Suitability Some firms now using robots expected in advance that the robots might prove unsuitable for the task envisaged (8 per cent), would have no advantage over hard automation (6 per cent), or would give no advantage over existing equipment (4 per cent). In the event the proportions turned out to be 13 per cent, 8 per cent and 4 per cent respectively, but none of them was among the six most commonly encountered kinds of difficulty.

Nevertheless, it should not be supposed that robots are nearly always, or even more often than not, the most suitable solution. In the feasibility studies undertaken for potential users over half recommended against the use of robots. There are many potential applications where simpler pick-and-place machines, hard automation, or non-automated equipment may be a more appropriate solution. Many of the robots at present in use, although originally acquired largely for their flexibility, have been treated in practice as dedicated equipment once an effective application has been developed. In some cases robots in this position have fallen into disuse once their 'dedicated' job has been superseded.

Reliability Altogether 26 per cent of current robot users expected and experienced problems with reliability and maintenance. Many firms have had more unintended downtime than they expected, and 24 per cent have had 'frequent' or 'very frequent' downtime due to problems with the robot itself. However, 35 per cent of downtime was due to problems with associated equipment, 17 per cent was due to difficulties with other equipment upstream or downstream, and 20 per cent was due to unsatisfactory components or materials.

These problems are often associated with the firm's lack of specialist expertise, a difficulty expected by 25 per cent of users and actually experienced by 27 per cent of them. They are compounded by difficulties resulting from inadequate after-sales support, experienced by no less than 33 per cent of users, and the more damaging because it is a problem which was foreseen in advance by only 8 per cent of users. This latter appears to be a particularly intractable problem for plants in remote locations far from the supplier's base and often an insoluble one when the supplier firm has gone out of business or lost a robot franchise.

The prospect of breakdowns in complex and ill-understood equipment is a nightmare for production managers. The obvious solutions are to foresee the potential problems and provide for them, either by acquiring the necessary expertise in-house, or by making very firm arrangements with the supplier, but in practice either solution can be difficult to achieve completely. Clearly the development of as much in-house skill as possible in maintenance is a prudent course for any user to take.

Industrial relations Altogether 31 per cent of robot users expected to meet problems due to opposition from shopfloor workers or trade unions, but only 2 per cent have actually encountered it in practice, while 17 per cent say they have enjoyed *better* labour relations as a result of the introduction of robots.

Before the introduction of robots it appears that the attitude of the workers directly affected was favourable in 41 per cent of firms, and unfavourable in 10 per cent, but pre-introduction attitudes appear to have had little effect on subsequent performance. After the introduction of robots attitudes have *improved*, not worsened, with 71 per cent favourable and only 4 per cent unfavourable. Full and early consultation with those directly affected has been associated with change from passive acceptance to positive welcome, and is the recommended course. The rare cases of opposition after introduction have been associated with unsatisfactory production performance. This points to the importance of effective consultation and planning.

Employment The introduction of robots has brought lower labour costs to more than half the plants using them, but this has rarely involved redundancies. Altogether 60 per cent of plants introducing robots report no change in employment as a direct result, 8 per cent an increase and 25 per cent a decrease. This has resulted in a net direct decrease of only about 700 jobs nationally, equivalent to only about one job per user and per four or five robots. Where jobs have been shed it has been almost entirely in the form of natural wastage and redeployment of displaced workers to other jobs within the plant. This seems to be a necessary condition for acceptance of the robots by the workers affected, but one that so far has not usually proved difficult to meet. These findings are in line with those of the PSI reports *Chips* and *Jobs* (PSI, 1985) and *Microelectronics in Industry: Promise and Performance* (PSI, 1986). However, as noted in the latter report, the rate of job loss due to the use of microelectronics has recently been increasing steeply, and the plants with decreases in jobs due to the use of robots are mainly those with the longer-established installations. It must therefore be expected that the increase in the use of robots and other AMT which is necessary for our international compet-

Policy Implications

itiveness is likely to result in greater job loss over the years. This in turn would make it necessary to expand industrial training and retraining as robot systems and other AMT systems are more widely implemented.

Skills and training   The introduction of robots normally leads to fewer unskilled shopfloor jobs and more multi-skilled ones. It is essential to make provision for training of the workers concerned in the special skills needed to operate and maintain robot systems. There has usually been no problem in persuading people to be trained and training has been provided by various organisations, the most commonly used being the robot suppliers. In general, robot users have so far found the training arrangements reasonably satisfactory. However, there is no room for complacency. The increased use of robots and other AMT will require increased efforts in retraining displaced workers and in developing a multi-skilled workforce.

A much more difficult problem in many cases is the provision of higher level skills in the form of production engineers with specific robotics expertise. Many robot users have felt that they have little choice but to rely on the suppliers for this, and the latter's resources appear sometimes to have become overstretched. This problem is likely to remain intractable until more of these engineers are trained by universities and polytechnics.

Government support   The survey revealed considerable lack of awareness of some of the government support schemes among potential users. Government should continue to make every effort to publicise the assistance which is available, but users and non-users must play their part. To whatever extent government support is available for advanced manufacturing technology in general, or robots in particular, it will make good sense for current and prospective users of robots to make it their business to keep themselves aware of the kinds of support in operation and the conditions on which they are available, so that they may benefit from the help they are intended to receive.

Robot suppliers

Market orientation   Robot suppliers may be expected to respond to market forces by identifying their potential customers and their needs and seeking to meet them. They will be able to do this more effectively if they have better information about where their current and prospective markets lie: the numbers of users and prospective users of robots, and the number of additional robots planned, in specific sectors and types of organisation. It is hoped that this report will add substantially to the information available on these matters.

Advanced manufacturing technology   Robots are being planned and used as elements in advanced manufacturing systems: their use ranges from forming part of an island of automation to acting as an integrated element of a flexible manufacturing system (FMS). If suppliers wish to maximise the successful use of robots they need to present them in this way and customers need to be educated to see robots as part of the spectrum of AMT. Equipment should therefore be developed with industry-standard communications between elements of integrated systems in mind.

Turnkey installation   Complete turnkey packages are already popular with many customers and are likely to become the preferred solution for an increasing proportion of users as the market for robots spreads down from the large firms with strong in-house technical expertise to the smaller users more dependent on the help which can be provided by suppliers. Turnkey packages demand specific applications expertise from the supplier. There is scope for the development of modular robotic equipment by manufacturers which can be 'tailored' by firms specialising in building systems for paticular applications.

After-sales support   Inadequate after-sales support is the second most commonly cited difficulty experienced by robot users, and it could become more widespread still as more potential users with less technical expertise come into the market. Most suppliers already aim to provide help within 24 hours in the event of breakdown, but there can be difficulties in providing this to users in distant locations or at a price which users find reasonable. Both difficulties seem to be accentuated by the shortage of the highly trained people used in this work and by the current level of investment in robots by British industry. There would seem to be an urgent need to train more people to be able to work in after-sales support.

Reliability   Altogether 35 per cent of robot users see greater reliability from robot systems as a prerequisite for using them further. A related issue is the need to improve reliability so that

Policy Implications

breakdowns are rarer and service calls less frequent. There should be market gains for suppliers able to establish a reputation for more reliable products, but improving the reliability of the robot itself is only part of the need. Downtime due to faults in associated equipment is even more common than downtime due to breakdown of the robot itself, and there is also much downtime as a result of problems with other equipment upstream or downstream and due to deficiencies in components and materials. There is a need, therefore, for suppliers and users to improve their integration planning.

Costs The high costs of robot system development is the most frequently cited of all the difficulties experienced by robot users, and the high costs of the robots themselves are also one of the most common difficulties given by users. When asked what would most promote more widespread use of robots in the future, no less than 70 per cent of robot users say cheaper robots and 53 per cent cheaper associated equipment. There could be a much bigger market for any supplier able significantly to reduce the cost of the equipment and bring down the cost of development work.

Ease of use Lack of technical expertise is one of the difficulties most frequently encountered by robot users. As a result, when asked what would most promote better use of robots, 56 per cent ask for easier programming, 43 per cent for less need for special skills, 39 per cent for easier maintenance, and 19 per cent for easier installation. If robots can be designed which are easier to use and make less demand on some scarce skills, they should find a ready market.

Performance A third area where robot users ask for improvements in robots is in their performance, with 50 per cent wanting better sensors, 37 per cent greater speed, 35 per cent more intelligence, 31 per cent more accuracy, and 29 per cent greater versatility. This may not be easy to deliver since the cost, for example, of new vision systems is still very high, and most users are not prepared to pay much more in order to get the various benefits.

Government

General economic situation A suggestion often made by both users and suppliers of robots is that what is most needed for encouraging the use of both advanced manufacturing systems in general and robots in particular is not so much measures aimed at them specifically as measures to improve the general economic climate in which they operate.

Suppliers selling in a number of different countries claim that robot sales in Britain have been particularly weak compared with some of the countries with which we compete because the recession has been particularly long and deep in this country. With manufacturing production static, and a high exchange rate encouraging competition from imports and making exports difficult, there has been considerable pessimism in the attitudes of existing and prospective robot users, with little enthusiasm for the investment in modern productive equipment needed for competitive efficiency. Whatever the advantages of current economic policies may be, the general conditions have not made it attractive to invest in robots, or easy to raise the money to do so. Suppliers therefore suggest that economic policy changes designed to create a climate more favourable to industrial expansion in general would be more help than anything directed at robots as such. Altogether 41 per cent of the robot users in the survey cite an upturn in the economy as something which would be particularly helpful in enabling them to make effective use of robots in the future.

Interest rates One aspect of general economic policy which is especially often mentioned by both suppliers and users is the exceptionally high level of interest rates. These are felt to put British industry at a severe disadvantage relative to other industrial countries where interest rates are much lower, and this is particularly true for investment in equipment like robots which pay off only over a relatively long period. Specifically, high interest rates are considered to be a major reason for the enforcement of the very short payback periods required on industrial investment in many British companies. This approach to investment appraisal effectively rules out investment in robots which need a longer period to produce a good return.

Investment support If for any reason a much lower level in the rate of interest which has to be paid to finance investment in industry cannot be achieved, then there is a strong case for selective government support for investment in advanced manufacturing technology. This should be seen not as specially favourable treatment for this investment but as a policy to offset the heavy disadvantage at present imposed by high rates of interest.

Policy Implications

In the course of the interviews, robot users repeatedly emphasised the crucial importance of the grant support received under the recently ended scheme. This scheme effectively made the payback period for robots realistic and so furnished the indispensable means of enabling the robot projects to pass the stringent financial tests currently obtaining in most companies. They were emphatic that without the grants the projects would mostly not have been undertaken, or if they had been undertaken they would not have progressed as far.

The same message was brought out in the postal survey, with half the companies in receipt of grants saying they would not have gone ahead at all without them, and nearly half of them saying that, even knowing what they know now, they would not be prepared to undertake a similar project without grant support. It may reasonably be inferred that potential new users, without the benefit of the experience acquired by existing users, would be even less likely to undertake expensive and difficult new investment of this kind. This view is supported by the comments made by prospective users in the interviews.

If support is given, it could take the form of interest relief grants, or it could take the form of tax allowances, the form often favoured by the larger companies — and a substantial proportion of existing robot users are the larger companies. However, if the object of policy is to encourage the spread of this new technology further down among the medium and smaller enterprises and plants, then cash grants on lines similar to those employed in the previous scheme would seem to be more appropriate. The benefit from these is more direct to calculate and more speedy to arrive, and so has a much more powerful incentive effect on the smaller companies with modest accounting sophistication and limited financial resources.

The differential in interest rates is currently so large that the recent support rate of 20 per cent would be unlikely to be sufficient to do the trick — indeed the lower rate of support has probably been a significant factor in the recent fall-off in robot sales. The earlier rate of 33.3 per cent would probably be more appropriate. For best effect it would be important to keep the procedures as simple as possible and make the processing as prompt as possible.

Awareness and demonstration   If there is neither a major fall in industrial interest rates nor substantial direct support, then the prospects of spreading advanced manufacturing technology throughout British industry are gloomy indeed. There are still a number of things which could usefully be done, however, and at relatively modest cost, in the way of promoting awareness and demonstration projects.

Many companies expressed a desire for an independent centre which could give advice and information for users and potential users, and which could arrange demonstrations and links between users and non-users. The Department of Trade and Industry's AMT programme has provided a great deal of useful information for industry and arranged visits to 'demonstration' firms. The impact of this could be enhanced by more publicity to ensure that far more firms are made aware of what is already available, and possibly by the establishment of regional AMT information centres to make the service more readily available across the country.

Feasibility studies   The support provided for feasibility studies has been widely used and, in general, to good effect. There is much approval in industry for this form of support and it needs to be continued. There has, however, been criticism of some of the consultants used; and analysis of the survey data shows that, compared with those who had feasibility studies done in other ways, the firms who had feasibility studies undertaken by consultants had more difficulties with all aspects of robot system development and integration than expected. Also, fewer projects made a contribution to profitability and more of them proved not worthwhile. Thus fewer firms with these problems decided to continue to invest in robots. In some instances the differences were marginal, and there were doubtless many individual cases where the use of consultants was a success. However, there are grounds for supposing that some of the consultants may not have had the specific expertise needed for best results on some of the studies on which they were employed. There may, therefore, be a case for more vetting of the consultants used.

Education   The shortage of production engineers with expertise in robots in particular and advanced manufacturing technology in general is a critical point of weakness. More are urgently needed. It is therefore vital to ensure that any further cuts in higher education do not take a form which curtail or reduce the quality of the courses currently available, and it will be important to do as much as possible to *expand* the provision of suitable course places in the future and to ensure that resources are available to train high quality AMT production engineers.

Policy Implications

Shopfloor training for robot use will be organised by companies. However, if increasingly large numbers of people are to be trained for multi-skilled flexibility, we cannot afford to fall behind countries like West Germany and Japan in the proportion of people who stay in education and training up to the age of 18. New initiatives will certainly be needed in this area.

One of the things that came over strongly in our talks with many different kinds of people was that, while a great variety of views were expressed on many different issues, very few suggested that government should not be active in encouraging investment in robot systems and other forms of advanced manufacturing technology. On the contrary, there seems to be a striking unanimity of view in industry that robots and other advanced technology applications form an area of special importance which gives rise to special problems; and that, while most of the action has to come from the robot users and the suppliers in industry itself, there are a number of key things that government can do and must do to enable them to get on with the job.

APPENDIX I   PARTICULARS OF THE STUDY

Form and purpose of the study

The study was undertaken by PSI on the initiative of the British Robot Association (BRA) and was funded by the Gatsby Charitable Foundation, the Department of Trade and Industry (DTI) and the National Economic Development Office (NEDO).

The purpose was to establish the basic facts of the use of industrial robots in Britain and to identify the factors affecting their use, including the difficulties and benefits expected and experienced and the impact of the support provided by the DTI, with a view to formulating proposals for policy initiatives to ensure fuller exploitation of the opportunities available.

The study has taken the form mainly of two concurrent postal surveys, one of existing robot users and the other of non-users with potential for becoming users in the future, in order to establish the basic facts, followed by case-study interviews with a limited number of users, non-users, suppliers and others, in order to gather qualitative information on the underlying issues.

Timing

The study was started in May 1985, with the summer and autumn used for a review of the information already available on the use of industrial robots in Britain and the for the preparation of the sample for the postal surveys. The survey questionnaires were despatched at the end of 1985 and the beginning of 1986 and the replies were analysed by computer in the spring of 1986. The complementary case-study interviews were also undertaken in the spring of 1986 and the report was prepared in the summer, printed in September and published in November.

Questionnaires

Those in the sample who were known in advance to be robot users were sent a six page yellow questionnaire with questions covering company particulars, introduction and use of robots, difficulties and benefits expected and experienced, industrial relations, employment and training, degrees of success with robots and plans for the future. Those who had received grants from the DTI were sent a green questionnaire, identical to the yellow one except that it included in addition a few further questions about the operation of the government's support scheme. The non-users, who were expected to have less information available, and to be less inclined to answer many questions, were sent a shorter three page pink questionnaire concerned mainly with their plans and expectations.

From the way the samples had to be compiled, for many firms it could not be known in advance whether they were already robot users or not. Accordingly they were sent both the yellow and the pink forms and invited to complete whichever was appropriate to their circumstances. In addition, a mini-questionnaire, covering only number of robots in use or planned, area of applications and place, size and industry, was sent to all those who had not responded to the initial questionnaire or the reminder sent out subsequently.

Appendix I   Particulars of the Study

Robots users sample

There were considerable problems involved in finding a satisfactory sample of robot users. It was known from the PSI survey of microelectronic applications in industry that only about one factory in 40 was actually using robots, so it would not be practical to survey a random sample of factories in order to get information from the few among them which were robot users. At the same time it was not straightforward to take a sample composed only of a representative group of robot users because there was not available any listing of all the robot users, nor even any precise knowledge of their total number.

It was therefore decided to build up a sample from more than one source. The DTI provided a list of those which had received grants under the robots support scheme. This gave 100 per cent coverage of the grant recipients but it was known that these accounted for a minority of all robot users.

It was therefore suggested by the BRA that the robot suppliers (who already provided particulars of the the numbers of robots supplied by them for the BRA's own annual returns) would be willing, on a confidential basis, to provide lists of their customers which, taken together, would account for virtually all the robot users. Initial soundings elicited a favourable response from suppliers, but the response to the formal request for addresses for the sample was very disappointing. While 15 of the 54 suppliers approached felt able to cooperate and provided lists which were useful and much appreciated, the remainder, in the event, did not.

It was therefore decided to supplement the sample by drawing on the respondents to PSI's survey of microelectronics in industry a year before who at that time had said they were using robots.

There were, of course, many duplications due to overlaps between the various lists and these were checked and removed before the questionaires were despatched. After deduction also of a few cases which could not be used because of change of address or going out of business, or which proved unsuitable for other reasons, the total list based on the grant recipients amounted to 171 addresses, that from the earlier PSI survey a further 65 addresses, and the robot suppliers a further 62, making a total of 298 altogether, as set out in Table A. In addition, further robot users were provided from the lists of potential users described below.

Non-users sample

There were also problems in finding an appropriate sample of potential users and it was decided that the best course would be to base one on the visitors to the Automan exhibition at the 1985 Advanced Manufacturing Summit at the National Exhibition Centre in Birmingham. Of the total of 21,414 people visiting the exhibition, those from outside the United Kingdom, those not interested in industrial robots and those with job titles suggesting they would not be appropriate for the survey were excluded, resulting in a total list of 11,603 names and addresses supplied by Cahners Exhibitions Ltd, the exhibition organisers. From this list were deleted:

- duplications with the DTI list,
- duplications with the PSI list,
- duplications with the suppliers' lists,
- unsuitable names or addresses (for example, non-commercial visitors, companies not in industry),
- multiple attendances (where there were several people from the same company, only one was retained, the selection being based on a specified hierarchy of job titles in terms of suitability for the survey).

Of the names and addresses remaining, every third one (in alphabetical order of companies) was taken, to give a list of 952 names and addresses actually used in the survey. This source was invaluable in providing the great majority of the addresses for the survey of non-users with potential for using robots, as well as useful additional numbers of existing users for inclusion in the survey of users.

It was decided to supplement the Automan list with those who in the 1985 PSI survey of microelectronics applications in industry had said they expected to be using robots in the next two years (a further 59 addresses), and also with companies from the BRA's list of members

Appendix I   Particulars of the Study

which, after deduction of some known to be neither users nor potential users, provided a further 74 addresses.

Thus the three sources together provided a total of 1,085 possible users, the majority of whom, as expected, were non-users, but about one fifth of whom turned out to be robot users already, thereby adding usefully to the total numbers in the users sample.

Despatch and response

The first batch of questionnaires was despatched on 29 November 1985 and followed by a reminder on 10 January 1986. The questionnaires for the addresses in the supplementary PSI and BRA lists were mostly sent out on 24 January 1986, with reminders three weeks later. The mini-questionnaires were sent to those who have not replied by early March.

The questionnaire forms were deliberately kept short and simple in the hope of getting a high response, and in the event the response rate for the known user sample was 57 per cent (or 73 per cent including the mini-questionnaires), which is very high for a survey of this kind. With the potential users the response of 41 per cent (53 per cent including the mini-questionnaires) was not quite so high, but was still high for a survey of this kind, particularly when it is borne in mind that the lists probably included a proportion of unsuitable addresses so that the true response rate was higher than the rate recorded. Particulars of the response rates of the various components of the samples are given in Table A.

The response was also good in the sense that nearly all the respondents answered the questions appropriately and only a very small percentage of the questions were left unanswered.

While the different component groups in the sample of users vary substantially, taken together the sample is almost identical in terms of plant size distribution with the profile of robot users provided by the PSI industry survey, and it is also very similar in terms of industrial composition. It thus seems likely that the sample used for the survey is free from serious bias in these respects. This, together with the high response rate, the substantial proportion of all users included (probably about one third), and the quality of the response to the questions, suggests that the information from the survey of robot users is likely to be a reliable guide to robot users as a whole.

Total numbers of robot users and robots in UK

For maximum usefulness the results of a sample survey need to be translated into terms of actual numbers in the country as a whole. With robots there are considerable difficulties in this because of the absence of comprehensive national data to which a sample can be related. Nevertheless, by combining the robots surveys findings with data from other sources, it is possible to arrive at working approximations of sufficient reliability to be useful.

The PSI survey of microelectronics in industry, which was based on a sample of 1,200 factories, closely controlled to be representative of the whole of UK manufacturing industry, indicated a total of about 560 robot users in Britain as a whole at the beginning of 1985. At the same time other factories expected to become users within the next two years, and the response of those of them included in the robots survey one year later, suggests that their rate of adoption has in fact been substantially less than they earlier expected, indicating a total of about 770 robot users altogether in the UK at the beginning of 1986.

Probably a better indicator (because it is based on bigger numbers) of the increase in the number of users since the time of the PSI industry survey is the proportion of current users in the robots survey who say they did not get their first robot until 1985. This suggests an increase of 31 per cent in the total number of robot users between the time of the industry survey at the beginning of 1985 and the robots survey at the beginning of 1986, indicating a UK total of about 770 robot users at the time of the latter.

Since the list of robot users provided by the DTI and the lists from other sources are independent of one another, the degree of overlap between them provides three further bases for estimating the total number of robot users. The overlap between the DTI list of robot users and the PSI one suggests a UK total of about 740 users; the overlap with the robot suppliers' lists a

Appendix I   Particulars of the Study

UK total of about 570; and the overlap with the list based on visitors to the Automan exhibition a UK total of about 840.

Of these five bases for estimates, the second one, combining the 1985 PSI industry survey with the increase since then shown by the 1986 robots survey, is probably the most likely to be accurate. Moreover, the figure of 740 produced by it is the same as that provided by one of the other methods, close to another of the others (770), and also very close to the mean of all five combined (730). It therefore seems a reasonable figure to adopt as a working assumption on the basis of the information presently available.

The robot users in the survey have an average of 4.3 robots each, and this figure seems compatible with the average numbers current and expected in the 1985 PSI industry survey.

The total number of robots at the end of 1985 is put at 3,208 in the BRA's annual publication *Robot Facts*. This compares with a figure of about 3,500 derived from the numbers in the PSI industry survey for early 1985 plus half the increase expected at that time in the next two years, or a figure of perhaps only about 2,500 robots if the increase in 1985 is heavily scaled down to allow for the much smaller rate of take-up which appears in fact to have taken place. The PSI figures are based on the stock of robots a year before and estimates of the increase in numbers since, while the BRA ones are based on the cumulative increases in numbers over a period of years, and both methods have limitations. Since, however, the BRA figures is already widely used, and there appears no strong reason for preferring any particular alternative figure, the most practical course, on the basis of information at present available, would seem to be to adopt it for the purpose of the calculations in the survey.

Thus the figures for the three elements — 740 robot users, an average of 3.4 robots each, and 3,208 robots in total — fit together and seem to constitute practical working assumptions for the total numbers in UK as a whole. While none of the three figures is likely to be exactly right, none of them is likely to be wrong by a very wide margin: the total number of users is probably correct to within +10 per cent and the total number of robots to within +20 per cent.

Future numbers of robot users

In the tables showing growth in total number of robots and robot users in the UK by 1988 (Tables 151 and 152), the high figures for the number of additional users are based on the assumptions that all the non-users in the sample who said they had plans to start using robots in the next two years will in fact do so (but that none of the other non-users will), that the plants in the non-users sample represented about one third of all the potential users visiting Automan, and that these in turn represented about half of all the potential users in UK. The low figures for the number of additional robot users (one fifth of the high ones) are based on the assumption that *all* potential users visited Automan and that only 40 per cent of those with plans to use robots will in fact do so (which is roughly the proportion of fall-off experienced in 1985).

The high figures for the additional number of robots which will be used by existing robot users by the beginning of 1988 are based on the numbers given by the respondents in the survey, scaled up to allow for the proportion of all UK users accounted for by the users in the sample. The low figures (one third of the high ones) seek to allow for the consideration that some may not implement their plans in full, or at all, and some of those who did not return the questionnaires may have less ambitious plans than those who did.

The high figures for the additional number of robots that will be used in plants which are at present non-users are based on the figures provided by the non-users in the survey with plans to use robots in the next two years, scaled up in line with the total number of new users on the assumptions described above. The low figures (one fifth of the high ones) are also scaled down in line with the assumptions above.

It will be evident that, compared with the figures for the UK at the time of the survey, the figures for two years later are based on a series of additional assumptions which inevitably are subject to much greater degrees of uncertainty. Hence there is a wide spread between the high figures and the low ones and a need to treat them as broadly indicative only.

Appendix I   Particulars of the Study

Case study interviews

The postal surveys were supplemented by a programme of case-study interviews with 10 robot users who had received DTI grant support for robots, 12 other users, 9 non-users with plans for robot use, and 4 robot suppliers. The users and prospective users were selected to encompass a cross-section of characteristics and experience.

The interviews were free-ranging and designed to get mainly qualitative informative to complement the basic factual information from the postal surveys, particularly on attitudes, motives and intentions and on the more sensitive and intangible issues not fully covered by the postal questionnaires. They were undertaken after the initial analysis of the postal survey had been completed so that they could focus more sharply on the issue of most interest, and follow up in more detail points already made by the respondents concerned in the postal questionnaires.

Presentation

The results of the postal surveys are presented mainly in the form of a series of tables which follow the main text of the report. The conventions and definitions used in the tables are explained in the notes immediately before them.

APPENDIX II: POSTAL SURVEY QUESTIONNAIRES
Standard questionnaire for robot users

(yellow)

IN CONFIDENCE

Industrial Robots Project

ROBOT USERS
If you have one or more industrial robots please complete this form

You will see that most of the questions can be answered simply by putting a tick in one of the boxes provided. If you would like to add further points of comment or explanation, please do so.

Where exact information is not readily available, please give approximate answers. If some questions cannot without difficulty be answered at all, please leave them blank and return the form anyway - *the other answers will still be of value.*

If you have any queries, please do not hesitate to contact Annette Walling, Colin Brown or Jim Northcott at PSI at the address above.

If you would like a summary of the findings of this survey, please tick this box. ☐

The questionnaire has been designed to take very little time to complete, and we hope that you will return it, in the reply paid envelope, as soon as possible.

MANY THANKS IN ANTICIPATION OF YOUR HELP

Appendix II   Postal Survey Questionnaires (yellow)

COMPANY PARTICULARS

1. **Is your company:**
   - a UK owned company? ☐
   - part of a UK owned group? ☐
   - part of an overseas owned group? ☐
   - a public sector enterprise? ☐

2. **How many plants does your company have in the UK?**
   - one plant ☐
   - 2 plants ☐
   - 3 - 5 plants ☐
   - more than 5 plants ☐

3. **Approximately how many people are employed at your plant at present?**
   - 1 - 19 ☐
   - 20 - 49 ☐
   - 50 - 99 ☐
   - 100 - 199 ☐
   - 200 - 499 ☐
   - 500 - 999 ☐
   - 1,000 - 4,999 ☐
   - 5,000 + ☐

4. **How many shifts a day do you have in the sections where you use (or plan to use) robots?**
   - one shift ☐
   - two shifts ☐
   - three shifts ☐
   - other *(please specify)* ☐

5. **Approximately what is your company's annual turnover at present?**
   - under £10,000 ☐
   - £10,000 - 99,000 ☐
   - £100,000 - 999,000 ☐
   - £1 million - 9 million ☐
   - £10 million - 99 million ☐
   - £100 million + ☐

6. **In which industry group is your company's main activity?**
   - food, drink, tobacco ☐
   - chemical, metals ☐
   - mechanical engineering ☐
   - electrical, instrument engineering ☐
   - vehicles, aircraft, shipbuilding ☐
   - other metal goods ☐
   - textiles ☐
   - clothing, footwear, leather, fur ☐
   - paper, printing, publishing ☐
   - bricks, cement, pottery, glass, wood products ☐
   - plastic products ☐
   - other manufacturing *(please specify)* ☐
   - non-manufacturing *(please specify)* ☐

7. **Does your company make or supply robots or associated equipment?**
   - yes ☐   no ☐

Appendix II Postal Survey Questionnaires   (yellow)

INTRODUCTION OF ROBOTS

8. Why did you first consider using industrial robots in your plant?

9. Where did the idea of using robots in your plant originate? And where was the final decision to acquire one taken?

	original idea	final decision
head office of group	☐	☐
company board of directors	☐	☐
plant management, below company board level	☐	☐
departmental level	☐	☐

10. Was a specific feasibility study undertaken before the decision was made to buy a robot?

    no ☐
    yes, using the robot supplier ☐
    yes, with help from company in group ☐
    yes, using in-house staff ☐
    yes, using an independent consultant ☐

11. If you used a consultant for a fesibility study, how did you rate the consultant's report?

    excellent ☐   good ☐   fair ☐   poor ☐   useless ☐

12. Were you aware that government grants are available for studies of the feasibility of using robots? If so, did you apply for a grant? And did you get one?

    not aware grant available ☐
    aware of availability, but did not apply for grant ☐
    applied for grant, but application not accepted ☐
    received grant for study ☐

13. At what stage is your present use of your first robot?

    experimental, under development, pre-production ☐
    installed for commercial production ☐
    abandoned, permanently out of use or sold ☐

14. When did your first acquire a robot?
    Please give year, and also month if in 1984 or 1985.

    _____

15. How long did it take to get it into commercial production?
    If it is not yet in production, please give number of months from when robot was delivered to expected start of production.

    _____

16. What approach to installation did you adopt with your first robot?

    complete turnkey installation handled by manufacturer or agent ☐
    complete turnkey installation handled by consultant or systems house ☐
    naked robot from supplier, installation planned and executed in-house ☐
    other *(please specify)* ☐

Appendix II Postal Survey Questionnaires (yellow)

USE OF ROBOTS

17. Have you more than one robot at your plant now? If so, how many?

　　　1 robot ☐　　2 robots ☐　　3 or more (please give number) _____

18. Approximately how much did your first robot cost? And your latest robot if you have more than one?
(Cost of robot itself, excluding associated equipment, upstream changes, development costs, etc.)

	first robot	latest robot
under £10,000	☐	☐
£10,000 - 19,000	☐	☐
£20,000 - 29,000	☐	☐
£30,000 - 39,000	☐	☐
£40,000 - 49,000	☐	☐
£50,000 +	☐	☐

19. In which industrial activities are you currently using a robot?
If you are using more than one robot, please put the number of robots beside the box. If a robot is performing more than one task, please enter it against each application.

arc welding	☐	drilling, riveting	☐	food, drink processing	☐
spot welding	☐	grinding, deburring	☐	textile handling	☐
painting, coating	☐	injection moulding	☐	cleaning	☐
glueing, sealing	☐	press loading	☐	inspection, testing	☐
assembly	☐	other machine loading	☐	training, eduction	☐
handling, palletising	☐	laser cutting	☐	other *(please specify)*	☐
packaging	☐	water jet cutting	☐		
casting	☐	heat treatment	☐		
forging	☐	wood working	☐		

20. For approximately how many hours a week is your robot scheduled to be running on commercial production?
If you have more than one robot in production, running different lengths of time, please try to give separate figures with number of robots on each.

　　　approx. number of hours a week. _____

21. Has the down time been more or less than expected?

　　　more ☐　　same ☐　　less ☐

22. What have been the main causes of unintended down time?

	very frequent	quite frequent	occasional	never
problems with robot itself	☐	☐	☐	☐
problems with associated equipment (attachments, guarding, feeders, storage, etc)	☐	☐	☐	☐
problems with upstream or downstream equipment or processes	☐	☐	☐	☐
problems with unsatisfactory components or materials	☐	☐	☐	☐
other *(please specify)*	☐	☐	☐	☐

Appendix II   Postal Survey Questionnaires
(yellow)

RESULTS OF USE OF ROBOTS

23. What benefits did you expect to get from the use of robots before going into production? And what ones have you actually experienced after going into production?
Please tick as many as apply.

	expected before	experienced after
improved quality, more consistent products	☐	☐
better management control	☐	☐
greater reliability, less down time	☐	☐
greater volume of output	☐	☐
greater flexibility for product changes	☐	☐
lower equipment costs	☐	☐
less capital tied up in work in progress	☐	☐
lower material costs, less waste	☐	☐
lower energy costs	☐	☐
lower labour costs	☐	☐
improved work conditions, environment, safety	☐	☐
better labour relations	☐	☐
increased technical expertise	☐	☐
other *(please specify)*	☐	☐

24. What difficulties or disadvantages did you expect from the use of robots before going into production? And what ones have you actually experienced after going into production?
Please tick as many as apply.

	expected before	experienced after
unsuitability of robot for task required	☐	☐
no advantage over existing equipment	☐	☐
no advantage over hard automation	☐	☐
insufficient reliability, maintenance problems	☐	☐
problems with installation and integration	☐	☐
inadequate after-sales support from suppliers	☐	☐
high costs of equipment	☐	☐
high costs of development	☐	☐
lack of specialist technical expertise	☐	☐
health and safety problems	☐	☐
opposition from shopfloor workers or unions	☐	☐
opposition from top management	☐	☐
opposition from other groups in company	☐	☐
other *(please specify)*	☐	☐

25. All in all, do you regard your use of robots as having been:

very worthwhile ☐
fairly worthwhile ☐
marginal ☐
not really worthwhile ☐
not at all worthwhile ☐

26. Has the use of robots made your operations more profitable?

yes ☐   no ☐

Appendix II Postal Survey Questionnaires (yellow)

27. In general, what has been the attitude of the workers directly affected by the introduction of robots?

	before introduction	after introduction
very favourable	☐	☐
quite favourable	☐	☐
neutral	☐	☐
quite unfavourable	☐	☐
very unfavourable	☐	☐

28. Were those directly affected consulted in advance of the first use of robots?

   yes ☐   no ☐

29. Has the introduction of robots been the subject of specific negotiations with the trade unions?

   yes ☐   no ☐

30. What has been the direct effect of the introduction of robots on the number of people employed at your plant?

   increase ☐   approx. number of increase _____
   no change ☐
   decrease ☐   approx. number of decrease _____

31. Approximately how many people have been on training courses as a result of the introduction of robots? And how many are expected to go in the next two years? And by whom are/were the courses provided?

	number already	number next two years
training course by own staff	_____	_____
training courses by staff from elsewhere in company or group	_____	_____
training courses by robot supplier	_____	_____
training courses by specialist training organisation	_____	_____
other *(please specify)*	_____	_____

32. How satisfied are you with your existing training arrangements? What improvements would you like to see?

FUTURE PLANS

33. Do you plan to acquire more robots in the next two years? If so, approximately how many?

   yes ☐   approx. number _____
   no ☐

34. If you get more robots, are they likely to be more sophisticated than those you have already? Or less?

   more sophisticated ☐
   about same ☐
   less sophisticated ☐

Additional questions in questionnaire for robot users in receipt of support grants *(green)*

GOVERNMENT SUPPORT

36. **At what stage did you become aware of the possibility of a grant to help you use robots?**
    - before your feasibility study ☐
    - during your feasibility study ☐
    - after your feasibility study ☐

37. **How did you first become aware of the possibility of support?**
    - trade, technical press ☐
    - general press ☐
    - radio, television ☐
    - exhibitions, trade fairs, conferences, courses ☐
    - equipment suppliers ☐
    - consultants ☐
    - other companies in same group ☐
    - other sources *(please specify)* ☐

38. **In the absence of support, would you have:**
    Please tick more than one if appropriate.
    - gone ahead, but on a smaller scale? ☐
    - gone ahead, but later? ☐
    - not gone ahead at all? ☐

39. **Given the knowledge and experience now gained, would you be prepared to undertake a similar project without a grant?**
    - no ☐ any special reasons?
    - yes ☐

40. **What government measures do you think would be most useful in helping firms in industry to use robots?**

Appendix II Postal Survey Questionnaires
(yellow)

35. **What would be paricularly helpful to enable you to make effective use of robots in the future?**
    Please tick as many as apply.

greater speed ☐	less need for special skills ☐	
more accuracy ☐	easier maintenance ☐	
heavier payload ☐	better after-sales service ☐	
greater versatility ☐	more UK based manufacturers ☐	
better sensors ☐	lower purchase price of robot ☐	
more intelligence ☐	lower cost of associated equipment ☐	
easier programming ☐	lower operating costs ☐	
smaller, lighter ☐	more government support ☐	
easier installation ☐	easier finance for investment ☐	
quicker delivery ☐	an upturn in the economy ☐	
better reliability ☐	other *(please specify)* ☐	

GOVERNMENT SUPPORT

36. **At the time you acquired your first robot, were you aware that in some cases government support is available to help firms with the purchase and development costs of robots.**

    not aware support available ☐
    not eligible for support ☐
    did not need support ☐
    level of support too low ☐
    conditions for support too stringent ☐
    processing of applications too slow ☐
    other *(please specify)* ☐

37. **What government measures do you think would be most useful in helping firms in industry to use robots?**

GENERAL

38. **Have you any other comments or suggestions of what most needs to be done to help ensure the most effective use of industrial robots in Britain?**

THANK YOU

Questionnaire for non-users of robots *(pink)*

1. **Is your company:**
   - a UK owned company? ☐
   - part of a UK owned group? ☐
   - part of an overseas owned group? ☐
   - a public sector enterprise? ☐

2. **How many plants does your company have in the UK?**
   - one plant ☐
   - 2 plants ☐
   - 3 - 5 plants ☐
   - more than 5 plants ☐

3. **Approximately how many people are employed at your plant at present?**
   - 1 - 19 ☐
   - 20 - 49 ☐
   - 50 - 99 ☐
   - 100 - 199 ☐
   - 200 - 499 ☐
   - 500 - 999 ☐
   - 1,000 - 4,999 ☐
   - 5,000 + ☐

4. **How many shifts a day do you have in the sections where you use (or plan to use) robots?**
   - one shift ☐
   - two shifts ☐
   - three shifts ☐
   - other *(please specify)* ☐

5. **Approximately what is your company's annual turnover at present?**
   - under £10,000 ☐
   - £10,000 - 99,000 ☐
   - £100,000 - 999,000 ☐
   - £1 million - 9 million ☐
   - £10 million - 99 million ☐
   - £100 million + ☐

6. **In which industry group is your company's main activity?**
   - food, drink, tobacco ☐
   - chemical, metals ☐
   - mechanical engineering ☐
   - electrical, instrument engineering ☐
   - vehicles, aircraft, shipbuilding ☐
   - other metal goods ☐
   - textiles ☐
   - clothing, footwear, leather, fur ☐
   - paper, printing, publishing ☐
   - bricks, cement, pottery, glass, wood products ☐
   - plastic products ☐
   - other manufacturing *(please specify)* ☐
   - non-manufacturing *(please specify)* ☐

7. **Does your company make or supply robots or associated equipment?**
   - yes ☐   no ☐

Appendix II   Postal Survey Questionnaires (pink)

8. **Have you any plans for using industrial robots?**

    in next two years ☐   if so, approximately how many? _____
    later ☐
    no firm plans, but possibility of
    using robots under consideration ☐
    use of robots not being considered ☐

9. **Has a specific feasibility study of the scope for using robots in your plant been undertaken? Or is one planned?**

    feasibility study already completed ☐   if so, did it recommend use of robots? ☐ yes
    feasibility study in progress ☐   ☐ no
    feasibility study planned ☐
    no plan at present for a feasibility study ☐

10. **Do you know that government grants are available for studies of the feasibility of using robots? Have you applied for a grant? Do you intend to in the future?**

    not previously aware that grant is available ☐
    aware of availability, but have not applied ☐
    intend to apply for grant in future ☐
    applied for grant, but application not yet decided on ☐
    applied for grant, but application not accepted ☐
    received grant for feasibility study ☐

11. **Do you know that government grants are in some cases available to help firms invest in robots?**

    aware of grants towards cost of robot ☐
    aware of grants towards cost of associated equipment ☐
    aware of grants for development work to get robot into production ☐

12. **If you do use robots in the future, what are the most likely areas of application?**
    Please tick any that seem likely.   no possible area of application ☐

    arc welding ☐   drilling, riveting ☐   food, drink processing ☐
    spot welding ☐   grinding, deburring ☐   textile handling ☐
    painting, coating ☐   injection moulding ☐   cleaning ☐
    glueing, sealing ☐   press loading ☐   inspection, testing ☐
    assembly ☐   other machine loading ☐   training, education ☐
    handling, palletising ☐   laser cutting ☐   other *(please specify)* ☐
    packaging ☐   water jet cutting
    casting ☐   heat treatment
    forging ☐   wood working

13. **Are there any developments (for example in the characteristics of robots, in the arrangements for their provision, in the circumstances of your firm, in the state of the economy, or in the policies of the government) which would significantly increase the likelihood of your using robots?**

Appendix II Postal Survey Questionnaires
(pink)

14. **What benefits would you expect to get from using robots?**
    - improved quality, more consistent products ☐
    - better management control ☐
    - greater reliability, less down time ☐
    - greater volume of output ☐
    - greater flexibility for product changes ☐
    - lower equipment costs ☐
    - less capital tied up in work in progress ☐
    - lower material costs, less waste ☐
    - lower energy costs ☐
    - lower labour costs ☐
    - improved work conditions, environment, safety ☐
    - better labour relations ☐
    - increased technical expertise ☐
    - other *(please specify)* ☐

15. **What difficulties or disadvantages would you expect from using robots?**
    Please tick any you consider to be very important.
    - unsuitability of robot for task required ☐
    - no advantage over existing equipment ☐
    - no advantage over hard automation ☐
    - insufficient reliability, maintenance problems ☐
    - problems with installation and integration ☐
    - inadequate after sales support from suppliers ☐
    - high costs of equipment ☐
    - high costs of development ☐
    - high production costs ☐
    - lack of specialist technical expertise ☐
    - health and safety problems ☐
    - opposition from shopfloor workers or unions ☐
    - opposition from top management ☐
    - opposition from other groups in company ☐
    - other *(please specify)* ☐

16. **Have you any other comments or suggestions about the use of industrial robots in Britain?**

THANK YOU

Abbreviated questionnaire (white)

PLEASE FILL IN BOTH SECTIONS A AND B

A

Do you have an industrial robot?   YES   NO

IF YES   Have you more than one robot at your plant now? If so, how many?

1 robot ☐   2 robots ☐   3 or more (please give number) _____

Do you plan to acquire more robots in the next two years? If so, approximately how many?

yes ☐   approx. number _____
no ☐

In which industrial activities are you currently using a robot?

arc welding ☐	drilling, riveting ☐	food, drink processing ☐
spot welding ☐	grinding, deburring ☐	textile handling ☐
painting, coating ☐	injection moulding ☐	cleaning ☐
glueing, sealing ☐	press loading ☐	inspection, testing ☐
assembly ☐	other machine loading ☐	training, eduction ☐
handling, palletising ☐	laser cutting ☐	other *(please specify)* ☐
packaging ☐	water jet cutting ☐	
casting ☐	heat treatment ☐	
forging ☐	wood working ☐	

IF NO   Have you any plans for using industrial robots?

in next two years ☐   if so, approximately how many? _____
later ☐
no firm plans, but possibility of using robots under consideration ☐
use of robots not being considered ☐

B

Approximately how many people are employed at your plant at present?

1 - 19 ☐	200 - 499 ☐		
20 - 49 ☐	500 - 999 ☐		
50 - 99 ☐	1,000 - 4,999 ☐		
100 - 199 ☐	5,000 + ☐		

In which industry group is your company's main activity?

food, drink, tobacco ☐	paper, printing, publishing ☐
chemical, metals ☐	bricks, cement, pottery, glass, wood products ☐
mechanical engineering ☐	plastic products ☐
electrical, instrument engineering ☐	other manufacturing *(please specify)* ☐
vehicles, aircraft, shipbuilding ☐	
other metal goods ☐	non-manufacturing *(please specify)* ☐
textiles ☐	
clothing, footwear, leather, fur ☐	

APPENDIX III DEPARTMENT OF TRADE AND INDUSTRY SUPPORT FOR ROBOTS

* The Robotics Support Programme (RSP) was launched by the Department of Trade and Industry in 1982. The scheme operated alongside the programme of support for Flexible Manufacturing Systems (FMS).

* The scheme offered financial assistance in three areas. First, support was offered towards the cost of feasibility studies. Firms could employ independent consultants or robot suppliers to carry out studies to determine whether robots could make a cost-effective contribution to their manufacturing business. Up to 50 per cent of the study costs could be paid as a grant. Secondly, support was available towards the cost of purchase and development of robot systems. Support was initially provided at up to 33.3 per cent of total eligible costs. Thirdly, 25 per cent grants were available towards R & D costs for UK manufacturers of robots and associated equipment.

* The main conditions of the scheme were:
  — A company could only re-apply for support after receiving a grant if the type of robot application proposed were new to the company.
  — Feasibility study grants were not available for in-house investigations.

* The Robots scheme and the FMS scheme were 'frozen' by a moratorium on applications in November 1984. The moratorium was lifted in March 1985, but the level of funding for the purchase and development of robot systems was reduced to 20 per cent.

* After March 1985 the grants for robots and FMS were available within a general framework, namely the Department's Advanced Manufacturing Technology (AMT) Programme. This provided advice, information and training as well as financial support, and covered the whole range of computer-based manufacturing technologies: robotics, FMS, CNC machine tools, CAD/CAM and so on. (See glossary in Appendix IV.)

* The AMT Programme accepted applications for grants towards purchase and installation costs up until June 1986. Since then DTI funding has been available through the long-running Support For Innovation scheme (SFI). This offers assistance towards the development of new products or of innovative processes, but not towards the purchase of equipment such as robots. In addition, regional development grants are still available to eligible companies. The AMT Programme continues to provide information on robotics and grants for feasibility/planning studies.

* In July 1985 the DTI launched an initiative in the field of Advanced Robotics (AR). AR is concerned with the development of highly sophisticated robots for use in hostile or extreme environments. Applications will include firefighting, tunnelling, emergency rescue work, and nuclear energy. The DTI scheme aims to promote AR research projects and the growth of a British AR industry.

APPENDIX IV GLOSSARY OF TERMS

This glossary provides definitions of selected terms currently in use in the field of manufacturing automation. We are grateful to Ingersoll Engineers for permission to draw on definitions given in their report *Integrated Manufacture* (IFS, 1985). This in turn draws on *How To Speak Automation*, published by General Electric in conjunction with Grant Publications in the USA.

AGV Automated guided vehicle. A vehicle equipped with automatic guidance facilities which follows a prescribed guide path to work areas for automatic or manual loading or unloading.

AMT Advanced manufacturing technology. General term to cover the various developing techniques in factory automation such as CAD, FMS and robotics.

Batch manufacture The production of parts in discrete runs or batches, interspersed with other production operations or runs of other parts.

CAD Computer-aided design. A CAD system allows the on-line construction of a highly detailed design drawing using a variety of interaction devices and programming techniques.

CAD/CAM Computer-aided design/computer-aided manufacture. This refers to the integration of computers into the entire design-to-fabrication cycle of a product or plant.

CAE Computer-aided engineering. This computer-based facility allows analysis of a design for basic error-checking or to optimise manufacturability, performance and economy. Information is drawn from a CAD/CAM design database.

CAM Computer-aided manufacturing. The use of computers to generate manufacturing data. Information from a CAD/CAM database can assist in or control a proportion or all of a manufacturing process. CAM techniques can be used to produce process plans for manufacturing a complete assembly; to program robots; and to coordinate plant operation.

CIM Computer-integrated manufacturing. The concept of a totally automated factory in which all manufacturing processes are integrated and controlled by a CAD/CAM system. CIM will enable production planners and schedulers, shop-floor foremen and accountants to use the same database as product designers and engineers.

CNC Computer numerical control. A technique by which a machine tool control uses a mini-computer to store NC instructions generated earlier by a CAD/CAM system for controlling the machine. See **DNC** and **NC**.

Dedicated equipment Equipment designed or intended for a single function or use.

DNC Direct numerical control. A system in which sets of NC machines are connected to a mainframe computer to establish a direct link between the DNC computer memory and the machine tools, which are directly controlled by the computer without the use of tape. See also **NC** and **CNC**.

Downtime The time when a system (such as a robot system) is not available for production due to breakdown in the hardware, software, ancillary equipment or associated processes and parts, or due to required maintenance.

Appendix IV Glossary of Technical Terms

FMC Flexible machining centre. Usually a multi-robot system that comprises CNC machines with robots loading and unloading parts that are conveyed into and through the system.

FMS Flexible manufacturing system. An arrangement of machines (typically machining centres under numerical control with tool changers) interconnected by a transport system. The transporter carries work to the machines on pallets or other units. A central computer controls machines and transport.

Gripper A robot 'hand' which picks up, holds and releases the part or object being handled.

Hard automation Use of specialised machines and machine lines to manufacture and assemble products or components. Hard automation is usually associated with continuous manufacturing and high volume production in distinction to batch manufacture.

Integrated system A CAD/CAM system which integrates the product development cycle (analysis, design and manufacture) so that all processes flow smoothly from concept to production.

Island of automation 'Stand-alone' automation products (robots, CAD/CAM systems, NC machines) without the integration required for a cohesive factory automation system.

MRP Materials requirements planning. A variety of computer applications for ordering materials and managing inventories.

Naked robot A robot which is bought 'off the shelf' and then developed, installed and integrated by the user.

NC Numerical control. A technique of operating machine tools or similar equipment in which motion is developed in response to numerically coded commands, usually stored on a tape. See also **CNC** and **DNC**.

Programming Robots can be programmed in the following ways. (1) The robot is placed in 'teach' mode of operation and points in space are recorded as the robot is led through the desired sequence of movements. (2) The robot, in teach mode, is manually walked through the sequence of movements and operations. (3) A pre-recorded program is transferred to the robot's control unit, usually via a magnetic tape. (4) Some robots are directly controlled from a computer running programs which specify the robots' movements and operations.

Robot The British Robot Association's definition is: a reprogrammable device designed to both manipulate and transport parts, tools or specialised manufacturing implements through variable programmed motions for the performance of specific manufacturing tasks.

Tool A term used loosely to define an attachment to a robot arm — for example, a simple gripper or an arc welding torch.

Turnkey system A robot system for which the supplier assumes total responsibility for building, installing and testing both hardware and software, and possibly also for training the user's staff.

WIP Work-in-progress. Products in various stages of completion throughout the manufacturing cycle, including raw material that has been released for initial processing and finished products awaiting final inspection and acceptance for shipment to a customer.

APPENDIX V   PSI'S MICROELECTRONICS APPLICATIONS RESEARCH PROGRAMME

Microelectronics in Industry: Promise and Performance

by Jim Northcott (June 1986) £29.95

A report based on a major survey of 1200 factories closely representative of the whole of manufacturing industry in Britain and two similar PSI surveys undertaken two and four years previously. It provides the most comprehensive and authoritative information currently available for measuring the changing pattern of development. Tables include breakdowns by type of application, size, industry, ownership and region, and figures weighted to represent all UK factories and share of total manufacturing employment. The tables are accompanied by a full commentary and interpretation, technical appendix, questionnaire, charts and summary.

Chips and Jobs: Acceptance of New Technology at Work

by Jim Northcott, Michael Fogarty and Malcolm Trevor (November 1985) £8.95

An appraisal of the available evidence of the impact of microelectronics technology on jobs and the changes in the nature of work, the effects on employment and the attitudes of the people affected.

Promoting Innovation: Microelectronics Applications Projects

by Jim Northcott with Simon Bennett, John Bessant, Steve Kelly, Richard Lamming, Liz Mills, Howard Rush and Sally Walker (September 1985) £10.00

An appraisal of the MAP scheme to support microelectronics development projects. The report, 175 pages with 107 tables, is based on Department of Trade and Industry records and PSI survey and case studies and examines the coverage, operation and effects of the scheme.

Microelectronics in British Industry: The Pattern of Change

by Jim Northcott and Petra Rogers (March 1984) £25.00

Report on the 1983 PSI survey of how far the 'microchip revolution' is actually taking place in Britain based on interviews with a representative nationwide sample of 1200 manufacturing establishments. It provides comprehensive and authoritative figures with breakdowns by size, ownership, industry, region and type of application and comparisons with a similar survey two years before. The 260 page report includes over 100 main tables, full particulars of the sample and survey method, a copy of the questionnaire and a brief, clear summary of the findings. Subjects covered include:

Form and extent of use	Obstacles and difficulties experienced
Stage of development reached	Use of government support schemes
Date production started	Skill shortages, recruitment and training
Type of Equipment used	Changes in jobs due to use of microelectronics
Performance relative to competitors	

Appendix V Other PSI Reports on Microelectronics Applications

Microelectronics in Industry: An International Comparison: Britain France and Germany

by Jim Northcott with Petra Rogers, Werner Knetsch and Berengere de Lestapis   (January 1985) £10.00

Report on the first major international comparison of its kind, based on the survey in Britain described above and parallel ones in France and Germany using a common questionnaire and similarly constructed samples of 1200 factories in each. It provides figures on a directly comparable basis for such things as overall extent of use, stage of development, breadth of diffusion, type of application, kind of equipment used, obstacles to adoption and changes in jobs. French and German language versions of the report are also available.

Microelectronics in Industry: What's Happening in Britain

by Jim Northcott with Petra Rogers   (March 1982) £5.00

Report on a survey undertaken two years earlier, covering the same subjects and based on the same sample of 1200 manufacturing establishments.

Microelectronics in Industry: Survey Statistics

by Jim Northcott and Petra Rogers with Anthony Zeilinger   (March 1982) £20.00

This supplementary volume gives a comprehensive set of over 200 tables from the survey, with breakdowns by type of application, industry, region and size of establishment, and also details of the sample and full particulars of the method and organisation of the survey.

Microelectronics in Industry: Interim Reports

by Jim Northcott with Petra Rogers and Anthony Zeilinger

1. **Extent of use**   (July 1981) £5.00
2. **Advantages and Problems**   (Sep. 1981) £5.00*
3. **Awareness and Government Support**   (Oct. 1981) £5.00
4. **Manpower and Training**   (Dec. 1981) £5.00*

Four interim reports on particular aspects of the survey described above.

* out of print

Microprocessors in Manufactured Products

by Jim Northcott with John Marti and Anthony Zeilinger   (November 1980) £3.25

The report is based on case studies of 90 companies in five different sectors offering potential scope for applications in their products — domestic electrical appliances, heating and ventilating equipment, cars, physical test instruments and toys and games. It examines the extent to which applications are being developed, the factors encouraging their adoption, the various problems and difficulties impeding progress, and the impact of the Government's MAP scheme to stimulate awareness, expand technical training, encourage the use of technical consultants, and support the cost of development work.

Microprocessors in Production Processes

by John Bessant   (July 1982) £5.00

The report is based on case studies of 130 companies, on similar lines to the study above, to find out from those taking the actual decisions in industry about the real issues affecting the introduction of microelectronics, but this time in the manufacturing processes by which products are made as opposed to in the products themselves.

Appendix V Other PSI Reports on Microelectronics Applications

Microprocessor Short Courses: Survey of Users

by Petra Rogers (July 1981) £5.0

The Government, through the agency of the National Computing Centre, has promoted the development of microprocessor short courses and achieved an increase of over 30,000 course places since 1978. How successful were they? With a view to finding out PSI undertook a survey of about 500 people who have been on these courses. It covers questions such as why they went on the courses, what they expected from them, how relevant they found them, how well they were organised, how they could be improved and whether they were likely to want further courses.

Micros and Money: New Technology in Banking and Shopping

by John Marti and Anthony Zeilinger (July 1982) £5.00

Microelectronics and other new technologies have potential applications in inter-bank payments systems, automated teller machines, transaction telephones, computerised stock control systems, direct electronic funds transfer systems at shop check-outs, view-data systems for shopping and banking from the home, plastic cards which can be 'charged' with a money value and which can be 'intelligent' through using their built-in chips to carry out their own transactions. The report examines the scope and cost-effectiveness of developments in prospect, and considers the implications for consumers, employees, retailers, bankers, and equipment makers.

The Electronic Office: Progress and Problems

by John Steffens (April 1983) £5.00

How far is the electronic 'office of the future' already taking shape today? PSI undertook a survey of 231 offices — all of them computer users — to establish the relative rates of adoption of different electronic products, the factors in choice of equipment, the benefits expected, the obstacles encountered, the attitudes of staff and the impact on jobs.

Copies of the Reports

PSI publications are obtainable from all good bookshops, or by visiting the Institute at 100, Park Village East, London NW1 3SR (01-387-2171).

Sales Representation: Frances Pinter (Publishers) Ltd. Orders to Martson Book Services, P O Box 87, Oxford, OX4 1LB.

TABLES

Presentation and definitions

The tables present the findings of two postal surveys carried out at the end of 1985 and beginning of 1986. They are based on samples comprising 248 robot users and 363 non-users, supplemented for a few questions by additional respondents bringing the sample numbers up to 326 users and 461 non-users.

The results are presented in terms of robot users, the sample of whom accounts for somewhere in the region of one third of all the robot users in Britain, and non-users, for whom the sample was selected mainly from visitors to the 1985 Automan exhibition. They may all be regarded as potential users; they are divided in the tables between those with definite plans to use robots within the next two years (i.e. by the beginning of 1988), those with no firm plans at present but who have the possibility of robot use under consideration, and those who are not currently contemplating robot use. For most of the questions breakdowns are provided by plant size, industry, stage of use and other relevant factors.

Details of the sample and other particulars of the survey are given in Appendix I and in Tables A–E which follow it.

Except where specified otherwise, the figures in the tables are for robot users and are mostly in the form of *percentages of establishments* in the samples giving the specified answers. The percentages refer to the BASE NUMBERS of plants given in *italics* at the top of the columns. The percentages are *rounded* to the nearest 1 per cent and consequently totals do not always add to 100. The figures for total number of robots and robot users in the UK are also rounded and therefore do not always add exactly to the totals shown.

The symbol 0 is used to represent zero or any percentage less than 0.5.

Although the figures are given mostly to the nearest 1 per cent for convenience of comparison, they should not be regarded as being necessarily *significant* to that degree of precision. All percentages should be read in relation to the base figures to which they refer, and it must be understood that where the base figures are small they can be regarded as representing only very broad orders of magnitude. Where the base figures is less than 20 the percentages are given in light type as a reminder that they should be regarded as very tentative.

Some figures are in terms of actual *numbers of robots, jobs, training course places* and so on, and these are given in *italics* to distinguish them from the percentages.

The figures in the *shaded areas refer not to the sample but to the whole of the UK*. They are *indicative only* and are based on the assumptions set out in Appendix I.

The *Q numbers* in the top left corners above the tables indicate the numbers of the questions in the questionnaire on which the tables are based, with the first number referring to the question being answered and the second number referring to the question providing the breakdown in terms of which the answer is being cross-tabulated. The NQ numbers refer to questions which are asked only in the non-users' questionnaire. The results are presented broadly in the order in which the questions occur in the questionnaires.

Tables

Tables 1–6, which set the robot users in the context of the rest of manufacturing industry, are derived from the PSI survey of microelectronics applications in industry which was carried out in 1985 and based on a sample of 1200 factories representative of British manufacturing as a whole. Full details of these figures are given in the PSI report *Microelectronics in Industry: Promise and Performance.*

For the purpose of the survey the British Robot Association's definition of a *robot* has been used: 'An industrial robot is a reprogrammable device designed to both manipulate and transport parts, tools or specialised manufacturing implements through variable programmed motions for the performance of specific manufacturing tasks', or, as the Department of Trade and Industry puts it more briefly, 'A robot is a reprogrammable mechanical manipulator'.

Robot users, for the purpose of the survey, are taken to include all those who have one or more than one industrial robot, including robots at the experimental or pre-production stage or in use for research, demonstrations or training within a manufacturing plant.

Table 1 Extent of use of robots compared with use of other kinds of automated manufacturing technology

users of robots and other kinds of equipment as a percentage of all UK factories*

Type of equipment	1985	1987 (expected in 1985)
Robots	1.6	4.8
Pick-and-place machines	3.2	8.0
Robots or pick-and-place machines or both	4.3	10.7
CNC machine tools	14	16
Programmable logic controllers	27	29
Any kind of microelectronically controlled production equipment	49	54

* From 'Microelectronics in Industry: Promise and Performance', report on PSI survey of microelectronics applications in industry in 1985, sample of 1200 weighted to represent all factories in UK manufacturing employing 20 or more people.

Table 2 Robot users' use of other advanced manufacturing technology

percentage of robot users and of all UK factories* using the technology

Type of technology	all factories	all microelectronics users	robot users
Integrated central control of groups of machines	6	11	19
Integrated central control of groups of processes	11	21	35
Integrated central control of machines or processes or both	13	25	40
Automated handling of products, materials or components	11	21	62
Automated storage	3	5	21
Computer aided design	12	22	42
Automated testing, quality control	20	37	66
Central machine/process control and automated handling and design and quality control	1	2	14

* From 'Microelectronics in Industry: Promise and Performance', report on PSI survey of microelectronics applications in industry in 1985, sample of 1200 weighted to represent all factories in UK manufacturing employing 20 or more people.

Table 3 Extent of use of robots by employment size of user

users of robots as a percentage of all UK factories* in size range

Number employed	1985	1987 (expected in 1985)
20–49	0	2
50–99	2	5
100–199	1	4
200–499	5	11
500–999	9	17
1000–	25	34
TOTAL	1.6	4.8

* From 'Microelectronics in Industry: Promise and Performance', report on PSI survey of microelectronics applications in industry in 1985, sample of 1200 composed of 200 in each size range representative of all UK factories in that size range.

Table 4 Extent of use of robots by industry

users of robots as a percentage of all UK factories* in industry

SIC class	Industry	1985	1987 (expected in 1985)
41/42	Food, drink, tobacco	0	1
25/26,21/22	Chemicals, metals	1	8
32	Mechanical engineering	2	6
33,34,37	Electrical, electronic, instrument engineering	4	5
35,36	Vehicles, aircraft, ships	8	17
31	Other metal goods	0	4
43	Textiles	1	3
44,45	Clothing, leather, footwear	0	1
47	Paper, printing, publishing	1	2
23/24,46,48,49	Other manufacturing	2	4
	TOTAL	1.6	4.8

* From 'Microelectronics in Industry: Promise and Performance', report on PSI survey of microelectronics applications in industry in 1985, sample of 1200 weighted to represent all factories in UK manufacturing employing 20 or more people.

Table 5 Extent of use of robots by type of company

users of robots as a percentage of all UK factories* in type of company

Type of Company	1985	1987 (expected in 1985)
UK independent	0.2	1.8
UK group	2.4	7.2
Overseas group	6.7	8.3
TOTAL	1.6	4.8

* From 'Microelectronics in Industry: Promise and Performance', report on PSI survey of microelectronics applications in industry in 1985, sample of 1200 weighted to represent all factories in UK manufacturing employing 20 or more people.

Table 6 Extent of use of robots by region

users of robots as a percentage of all UK factories* in region

Region	1985	1987 (expected in 1985)
Scotland	1	1
North	1	9
Yorkshire/Humberside	2	8
North West	3	5
East Midlands	2	5
West Midlands	2	4
East Anglia	1	4
South East	1	4
South West	0	4
Wales	2	2
TOTAL	1.6	4.8

* From 'Microelectronics in Industry: Promise and Performance', report on PSI survey of microelectronics applications in industry in 1985, sample of 1200 weighted to represent all factories in UK manufacturing employing 20 or more people.

Q.3 × Q.13

Table 7 Distribution of robot users by employment size and stage of use of first robot

column percentages

Number employed at plant	experimental, pre-production	installed for commercial production	abandoned or sold	TOTAL	Total number of users in UK
BASE	*48*	*182*	*16*	*248*	
1–19	8	1	0	2	*20*
20–49	4	5	0	4	*30*
50–99	0	8	0	6	*40*
100–199	15	11	0	11	*90*
200–499	21	29	38	27	*210*
500–999	23	14	25	17	*110*
1000–4999	23	23	19	23	*180*
5000–	6	9	19	9	*60*
Not answered	0	1	0	1	
TOTAL	100	100	100	100	*740*

Note: Figures for the total number of robot users in UK should be regarded as indicative only. Particulars of the assumptions on which they are based are given in Appendix I.

Q.5 × Q.13

Table 8 Distribution of robot users by company turnover and stage of use of first robot

column percentages

Annual turnover of company	experimental, pre-production	installed for commercial production	abandoned or sold	TOTAL	Total number of users in UK
BASE	*48*	*182*	*16*	*248*	
Under £10,000	0	0	0	0	*0*
£10,000–£99,000	4	0	0	1	*10*
£100,000–£999,000	6	3	0	4	*30*
£1 million–£9 million	21	28	13	25	*190*
£10 million–£99 million	35	37	56	38	*280*
£100 million–	25	26	31	26	*190*
Not answered	8	6	0	6	*40*
TOTAL	**100**	**100**	**100**	**100**	*740*

Note: Figures for the total number of robot users in UK should be regarded as broadly indicative only. Particulars of the assumptions on which they are based are given in Appendix I.

Q.2 × Q.13

Table 9 Distribution of robot users by number of plants and stage of use of first robot

column percentages

Number of UK plants	experimental, pre-production	installed for commercial production	abandoned or sold	TOTAL	Total number of users in UK
BASE	*48*	*182*	*16*	*248*	
1	33	31	19	30	*220*
2	8	17	19	16	*120*
3–5	23	19	6	19	*140*
5+	33	33	56	34	*250*
Not answered	2	1	0	1	
TOTAL	**100**	**100**	**100**	**100**	*740*

Note: Figures for the total number of robot users in UK should be regarded as broadly indicative only. Particulars of the assumptions on which they are based are given in Appendix I.

Q.6 × Q.13

Table 10 Distribution of robot users by industry and stage of use of first robot

column percentages

	experimental, pre-production	installed for commercial production	abandoned or sold	TOTAL	Total number of users in UK
BASE	48	182	16	248	
Food, drink, tobacco	2	3	6	3	30
Chemicals, metals	8	4	0	5	30
Mechanical engineering	27	24	6	23	180
Electrical, electronic, instrument engineering	17	13	19	15	110
Vehicles, aircraft, shipbuilding	15	24	25	22	160
Other metal goods	6	14	13	13	80
Textiles	2	0	0	0	0
Clothing, footwear, leather, fur	4	1	0	1	10
Paper, printing, publishing	0	3	0	2	10
Bricks, cement, pottery, glass, wood products	4	4	13	4	30
Plastic products	4	10	19	9	70
Other manufacturing	4	4	0	4	20
Non-manufacturing	8	1	0	2	10
Not answered	4	1	0	2	
TOTAL	106*	105*	100	105*	740

Note: Figures for the total number of robot users in UK should be regarded as broadly indicative only. Particulars of the assumptions on which they are based are given in Appendix I.

* Totals add to slightly more than 100% because a few respondents gave more than one industry for their company's main activity.

103

Q.1,2,3,5 × Q.6

Table 11 Distribution of robot users by industry group and by employment size, turnover and number of plants

column percentages

	mech. eng.	elec. eng.	vehicles aircraft ships	other metal goods	plastics	other	TOTAL
BASE	57	36	55	31	23	53	248
Number of employees at plant							
1–99	25	8	5	19	17	8	13
100–999	63	67	33	68	78	51	55
1000+	12	25	62	23	4	40	32
Annual turnover of company							
under £10 million	53	20	13	39	43	21	30
£10–£99 million	37	44	27	52	44	38	38
over £100 million	9	28	55	7	0	38	26
Number of plants in UK							
one plant only	42	31	16	42	39	19	30
two to four plants	39	28	31	29	30	42	34
more than five plants	19	42	53	29	30	38	34

Q.4 × Q.13

Table 12 Distribution of robot users by number of shifts worked and stage of use of first robot

column percentages

Number of shifts per day	experimental, pre-production	installed for commercial production	abandoned or sold	TOTAL	Total number of users in UK
BASE	48	182	16	248	
One	54	29	13	33	240
Two	25	45	31	40	300
Three	17	22	44	22	160
Other system	4	2	13	3	20
Not answered	0	2	0	2	20
TOTAL	100	100	100	100	740

Note: Figures for the total number of robot users in UK should be regarded as broadly indicative only. Particulars of the assumptions on which they are based are given in Appendix I.

Q.7 × Q.13

Table 13 Distribution of robot users between those which are and are not robot suppliers by stage of use of first robot

column percentages

	experimental, pre-production	installed for commercial production	abandoned or sold	TOTAL
BASE	48	182	16	248
Company makes or supplies robots or associated equipment:				
YES	21	8	19	11
NO	77	91	81	88
Not answered	2	1	0	1
TOTAL	100	100	100	100

Q.13,1

Table 14 Stage of use of first robot by type of company

column percentages

	UK company	overseas company	TOTAL
BASE	197	47	248
Experimental, under development, pre-production	21	13	19
Installed for commercial production	72	81	73
Abandoned, permanently out of use, or sold	7	6	7
Not answered	0	0	1
TOTAL	100	100	100

Q.13 × Q.14

Table 15 Stage of use of first robot by year first robot acquired

column percentages

	before 1981	1981–1982	1983	1984	1985	TOTAL	Total number of users in UK
BASE	*33*	*38*	*41*	*70*	*57*	*248*	
Experimental, under development, pre-production	0	8	12	17	46	19	*140*
Installed for commercial production	76	84	83	80	54	73	*540*
Abandoned, permanently out of use, or sold	21	8	5	3	0	7	*50*
Not answered	3	0	0	0	0	1	
TOTAL	100	100	100	100	100	100	*740*

Note: Figures for the total number of robot users in UK should be regarded as broadly indicative only. Particulars of the assumptions on which they are based are given in Appendix I.

Q.14 × Q.13

Table 16 Year first robot acquired by stage of use of first robot

column percentages

Year	experimental, pre-production	installed for commercial production	abandoned or sold	TOTAL
BASE	*48*	*182*	*16*	*248*
Before 1970	0	1	13	1
1970–1974	0	1	6	1
1975–1979	0	9	19	9
1980	0	3	6	3
1981	2	7	0	6
1982	4	10	19	10
1983	10	19	13	17
1984	25	31	13	28
1985	54	17	0	23
Not answered	4	2	13	4
TOTAL	100	100	100	100

Q.14 × Q.3

Table 17 Year first robot acquired by employment size of plant

cumulative column percentages, excluding not answered

Year	1–99	100–199	200–499	500–999	1000	TOTAL
BASE	29	26	66	40	77	239
Before 1975	0	0	5	3	1	2
1975–1979	0	0	20	8	13	11
1980	0	4	23	10	17	14
1981	3	4	24	20	26	20
1982	7	8	36	35	36	30
1983	24	38	50	45	56	47
1984	72	65	79	70	82	76
1985	100	100	100	100	100	100

Q.14 × Q.2

Table 18 Year first robot acquired by number of plants of company

cumulative column percentages, excluding not answered

Year	1	2	3–5	5–	TOTAL
BASE	72	38	44	83	239
Before 1975	3	3	0	2	2
1975–1979	8	11	11	13	11
1980	13	13	11	17	14
1981	15	21	14	25	20
1982	19	37	18	41	30
1983	39	47	45	54	47
1984	67	79	80	82	76
1985	100	100	100	100	100

Q.14 × Q.4

Table 19 Year first robot acquired by number of shifts a day

cumulative column percentages, excluding not answered

Year	one	two	three	TOTAL
BASE	79	96	53	239
Before 1975	1	2	2	2
1975–1979	5	13	17	11
1980	9	14	19	14
1981	10	20	26	20
1982	19	31	38	30
1983	37	49	55	47
1984	71	79	81	76
1985	100	100	100	100

Q.14 × Q.6

Table 20 Year first robot acquired by industry group

cumulative column percentages, excluding not answered

Year	mech. eng.	elec. eng.	vehicles aircraft ships	other metal goods	plastics	other	TOTAL
BASE	53	35	54	31	22	52	239
Before 1975	2	0	4	6	0	0	2
1975–1979	9	9	17	13	14	4	11
1980	9	17	19	19	18	4	14
1981	13	23	30	19	23	10	20
1982	25	31	39	29	23	23	30
1983	38	60	57	42	59	35	47
1984	75	74	85	68	82	71	76
1985	100	100	100	100	100	100	100

Q.9 × Q.13

Table 21 Origin of idea and place of decision to use robots by stage of use of first robot

column percentages

	experimental, pre-production	installed for commercial production	abandoned or sold	TOTAL	Total number of users in UK
BASE	48	182	16	248	
Origin of idea					
Head Office	2	8	13	7	50
Company board	19	30	19	27	200
Plant management	35	42	56	42	310
Department	31	19	13	21	160
Not answered	13	1	0	4	
TOTAL	100	100	100	100	740
Final Decision					
Head Office	10	17	13	15	120
Company board	58	60	56	59	450
Plant management	21	17	19	17	140
Department	2	3	0	3	20
Not answered	0	3	13	5	
TOTAL	100	100	100	100	740

Q.9 × Q.3

Table 22 Origin of idea and place of decision to use robots by employment size of plant

column percentages

	1–99	100–199	200–499	500–999	1000–	TOTAL
BASE	31	27	68	42	78	248
Origin of idea						
Head office	3	7	7	2	10	7
Company board	68	33	32	17	10	27
Plant management	26	44	49	38	42	42
Department	0	4	12	38	35	21
Not answered	3	11	0	5	3	4
TOTAL	100	100	100	100	100	100
Final decision						
Head office	10	19	18	14	15	15
Company board	84	70	62	55	47	59
Plant management	3	7	16	17	27	17
Department	0	0	2	5	5	3
Not answered	3	4	3	10	5	5
TOTAL	100	100	100	100	100	100

Q.9 × Q.2

Table 23 Origin of idea and place of decision to use robots by number of plants of company

column percentages

	1	2	3–5	5–	TOTAL
BASE	75	39	46	85	248
Origin of idea					
Head office	4	8	7	9	7
Company board	41	31	20	18	27
Plant management	40	44	48	38	42
Department	11	!5	24	32	21
Not answered	4	3	2	4	4
TOTAL	100	100	100	100	100
Final decision					
Head office	11	26	15	14	15
Company board	79	62	52	47	59
Plant management	7	8	26	26	17
Department	1	3	2	5	3
Not answered	3	3	4	8	5
TOTAL	100	100	100	100	100

Q.9 × Q.6

Table 24 Origin of idea and place of decision to use robots by industry group

column percentages

	mech. eng.	elec. eng.	vehicles aircraft ships	other metal goods	plastics	other	TOTAL
BASE	*57*	*36*	*55*	*31*	*23*	*53*	*248*
Origin of idea							
Head office	5	8	4	10	4	9	7
Company board	44	25	15	29	35	26	27
Plant management	37	39	51	45	48	30	42
Department	11	25	29	16	13	26	21
Not answered	4	3	2	0	0	8	4
TOTAL	100	100	100	100	100	100	100
Final decision							
Head office	16	11	18	13	4	19	15
Company board	70	56	56	77	70	42	59
Plant management	9	19	18	7	22	30	17
Department	2	3	4	0	4	4	3
Not answered	4	11	4	3	0	6	5
TOTAL	100	100	100	100	100	100	100

Q.9 × Q.17

Table 25 Origin of idea and place of decision to use robots by number of robots in use

column percentages of plants, mean numbers of robots per plant, approximate total numbers of robots in UK and percentages of total numbers of robots in UK

	PLANTS					ROBOTS	
	robots per plant			TOTAL	mean number of robots per plant	total number of robots in UK	per cent of total robots in UK
	1	2–5	6+				
	%	%	%	%	no.	no.	%
BASE	104	89	43	248			3208
Origin of ideas							
Head office	6	7	5	7	10	400	13
Company board	36	22	19	27	4	700	22
Plant management	39	42	40	42	4	1300	41
Department	14	26	30	21	4	700	22
Not answered	5	3	2	4	3	50	2
TOTAL	100	100	100	100	4.3	3208	100
Final decision							
Head office	17	16	5	15	2	250	7
Company board	66	56	51	59	5	1850	58
Plant management	12	18	30	17	7	850	26
Department	1	4	5	3	4	100	3
Not answered	4	6	9	5	5	200	6
TOTAL	100	100	100	100	4.3	3208	100

Note: Figures for the total number of robots in UK should be regarded as broadly indicative only. Particulars of the assumptions on which the figures are based are given in Appendix I.

Q.10 × Q.13

Table 26 Organisation undertaking feasibility study by stage of use of first robot

column percentages

Organisation undertaking Feasibility study	experimental, pre-production	installed for commercial production	abandoned or sold	TOTAL	Total number of users in UK
BASE	48	182	16	248	
Own company in-house	44	37	31	38	280
Other company in group	8	8	0	7	50
Robot supplier	25	32	25	30	220
Consultant	15	15	6	14	110
No study undertaken	8	8	25	10	80
Not answered	2	3	13	4	
TOTAL	102*	103*	100	103*	740

* Some totals add to more than 100 because a few plants had feasibility studies undertaken by more than one organisation.

Q.10 × Q.3

Table 27 Organisation undertaking feasibility study by employment size of plant

column percentages

	1–99	100–199	200–499	500–999	1000–	TOTAL
BASE	31	27	68	42	78	248
Own company in-house	29	37	34	38	45	38
Other company in group	3	11	3	7	10	7
Robot supplier	26	30	35	33	27	30
Consultant	19	15	18	14	9	14
No study undertaken	19	0	6	14	10	10
Not answered	3	7	6	0	1	4
TOTAL	100	100	102*	106*	102*	103*

* Some totals add to more than 100 because a few plants had feasibility studies undertaken by more than one organisation.

Q.10 × Q.17

Table 28 Organisation undertaking feasibility study by number of robots in use

column percentages of plants, mean numbers of robots per plant, approximate total numbers of robots in UK and percentages of total numbers of robots in UK

	PLANTS					ROBOTS	
	robots per plant				mean number of robots per plant	total number of robots in UK	per cent of total robots in UK
	1	2–5	6+	TOTAL			
	%	%	%	%	no.	no.	%
BASE	104	89	43	248			3208
Own company in-house	39	31	49	38	4	1550	48
Other company in group	8	7	5	7	4	150	5
Robot supplier	31	29	30	30	3	850	26
Consultant	17	15	5	14	6	300	10
No study undertaken	8	11	9	10	3	300	9
Not answered	2	7	2	4	3	100	3
TOTAL	105*	100	100	103*	4.3	3208	101*

* Some totals add to more than 100 because a few plants had feasibility studies undertaken by more than one organisation.

Q.11 × Q.13

Table 29   Rating of consultant's report by stage of use of first robot

column percentages

	experimental, pre-production	installed for commercial production	TOTAL
BASE	*6*	*24*	*31*
Excellent	17	4	7
Good	50	33	39
Fair	33	42	39
Poor	0	8	7
Useless	0	8	7
Not answered	0	4	3
TOTAL	*100*	*100*	*100*

Q.16 × Q.13

Table 30   Approach to installation of first robot by stage of use of first robot

column percentages

Approach to installation	experimental, pre-production	installed for commercial production	abandoned or sold	TOTAL
BASE	*48*	*182*	*16*	*248*
Turnkey by manufacturer	33	53	31	48
Turnkey by consultant	8	3	0	4
Naked robot	50	32	63	38
Other	8	10	0	9
Not answered	0	2	6	2
TOTAL	*100*	*100*	*100*	*100*

Q.16 × Q.10

Table 31 Approach to installation of first robot by type of feasibility study

column percentages

	in-house	company in group	robot supplier	consul-tant	none	TOTAL
BASE	*93*	*18*	*75*	*35*	*24*	*248*
Turnkey by manufacturer	42	56	57	43	42	48
Turnkey by consultant	1	0	3	17	4	4
Naked robot	46	22	31	31	50	38
Other	11	17	8	9	4	9
Not answered	0	6	1	0	0	2
TOTAL	**100**	**100**	**100**	**100**	**100**	**100**

Q.16 × Q.6

Table 32 Approach to installation of first robot by industry group

column percentages

	mech. eng.	elec. eng.	vehicles aircraft ships	other metal goods	plastics	other	TOTAL
BASE	*57*	*36*	*55*	*31*	*23*	*53*	*248*
Turnkey by manufacturer	53	28	60	61	44	38	48
Turnkey by consultant	2	8	2	0	9	6	4
Naked robot	35	53	31	23	39	45	38
Other	9	11	7	13	9	8	9
Not answered	2	0	0	3	0	4	2
TOTAL	**100**	**100**	**100**	**100**	**100**	**100**	**100**

Q.15 × Q.13

Table 33 Number of months needed to get first robot into commercial production by stage of use of first robot

column percentages and cumulative column percentages, excluding not answered

Number of months needed	experimental, pre-production		installed for commercial production	
	(expected)		(actual)	
	%	cum. %	%	cum. %
BASE	48	48	182	182
0	38	38	9	9
1–3	13	50	35	45
4–6	15	65	26	71
7–9	4	69	6	76
10–12	8	77	13	90
13–18	8	85	6	96
19–24	8	94	4	99
over 24	6	100	1	100
Not answered	38	–	0	–
TOTAL	100	100	100	100
Mean number of months	7.6	7.6	6.4	6.4

Q.15 × Q.8

Table 34 Number of months needed to get first robot into commercial production by type of feasibility study

column percentages of plants with first robot already in production

Number of months needed	in-house	company in group	robot supplier	consul-tant	none	TOTAL
BASE	67	14	59	27	15	182
0	3	14	14	22	7	9
1–3	45	36	24	41	27	35
4–6	28	43	32	7	13	26
7–9	8	0	3	7	7	6
10–12	12	0	14	11	27	13
over 12	4	7	14	11	20	10
TOTAL	100	100	100	100	100	100
Mean number of months	6	4	7	5	10	6.4

Q.15 × Q.6

Table 35 Number of months needed to get first robot into commercial production by industry group

column percentages of plants with first robot already in production

Number of months needed	mech. eng.	elec. eng.	vehicles aircraft ships	other metal goods	plastics	other	TOTAL
BASE	43	24	44	26	18	35	182
0	5	13	9	4	28	3	9
1–3	37	25	21	54	39	43	35
4–6	26	29	39	15	6	29	26
7–9	14	0	5	8	6	6	6
10–12	7	17	11	19	22	11	13
over 12	12	17	16	0	0	9	10
TOTAL	100	100	100	100	100	100	100
Mean number of months	7	7	7	5	4	6	6.4

Q.12

Table 36 Awareness of government grants for feasibility studies by whether received support for development of robot application

column percentages

	plants with grants to assist robot applications	other plants	TOTAL
BASE	*109*	*139*	*248*
Not aware grants available for studies of feasibility of using robots	10	22	17
Aware, but did not apply	41	48	45
Applied, but not accepted	0	12	7
Received feasibility study grant	31	9	19
Not answered	17	10	13
TOTAL	100	100	100

Q.12 × Q.10

Table 37 Awareness of government grants for feasibility studies by type of feasibility study

column percentages

	in-house	company in group	robot supplier	consul-tant	none	TOTAL
BASE	*93*	*18*	*75*	*35*	*24*	*248*
Not aware grants available for studies of feasibility of using robots	17	6	23	0	25	17
Aware, but did not apply	52	61	48	17	50	45
Applied, but not accepted	8	11	4	3	8	7
Received feasibility study grant	0	17	19	80	4	19
Not answered	24	6	7	0	13	13
TOTAL	100	100	100	100	100	100

Q.12 × Q.1

Table 38 Awareness of government grants for feasibility studies by type of company

column percentages

	UK company	overseas company	TOTAL
BASE	197	47	248
Not aware grants available for studies of feasibility of using robots	16	17	17
Aware, but did not apply	47	40	45
Applied, but not accepted	7	4	7
Received feasibility study grant	18	23	19
Not answered	13	15	13
TOTAL	100	100	100

Q.12 × Q.3

Table 39 Awareness of government grants for feasibility studies by employment size of plant

column percentages

	1–99	100–199	200–499	500–999	1000–	TOTAL
BASE	31	27	68	42	78	248
Not aware grants available for studies of feasibility of using robots	10	26	19	21	12	17
Aware, but did not apply	61	41	40	38	49	45
Applied, but not accepted	3	0	4	7	12	7
Received feasibility study grant	20	19	26	15	14	19
Not answered	6	14	10	19	14	13
TOTAL	100	100	100	100	100	100

Q.12 × Q.6

Table 40 Awareness of government grants for feasibility studies by industry group

column percentages

	mech. eng.	elec. eng.	vehicles aircraft ships	other metal goods	plastics	other	TOTAL
BASE	57	36	55	31	23	53	248
Not aware grants available for studies of feasibility of using robots	16	17	13	13	22	21	17
Aware, but did not apply	46	36	55	61	35	43	45
Applied, but not accepted	9	3	11	3	0	6	7
Received feasibility study grant	17	25	11	10	30	20	19
Not answered	13	20	11	13	13	10	13
TOTAL	100	100	100	100	100	100	100

Q.12 × Q.17

Table 41 Awareness of government grants for feasibility studies by number of robots in use

column percentages of plants, mean numbers of robots per plant, approximate total numbers of robots in UK and percentages of total numbers of robots in UK

	PLANTS					ROBOTS	
	robots per plant			TOTAL	mean number of robots per plant	total number of robots in UK	per cent of total robots in UK
	1	2–5	6+				
	%	%	%	%	no.	no.	%
BASE	104	89	43	248		3208	
Not aware grants available for studies of feasibility of using robots	18	11	21	17	4	500	16
Aware, but did not apply	48	45	40	45	3	1100	35
Applied, but not accepted	3	10	9	7	11	400	13
Received feasibility study grant	18	20	14	19	6	1050	33
Not answered	13	14	17	13	3	100	4
TOTAL	100	100	100	100	4.3	3208	100

Note: Figures for the total number of robots in UK should be regarded as broadly indicative only. Particulars of the assumptions on which they are based are given in Appendix I.

Q.36 × Q.3

Table 42 Time of awareness of support to help use robots of plants receiving grants by employment size of plant

column percentages

Stage first became aware of possibility of grant	1–99	100–199	200–499	500–999	1000–	TOTAL
BASE	20	13	35	13	28	110
Before feasibility study	75	54	83	77	71	75
During feasibility study	25	46	9	15	18	19
After feasibility study	0	0	3	0	4	2
Not answered	0	0	6	8	7	5
TOTAL	100	100	100	100	100	100

Q.36 × Q.6

Table 43 Time of awareness of support to help use robots of plants receiving grants by industry group

column percentages

Time first became aware of possibility of grant	mech. eng.	elec. eng.	vehicles aircraft ships	other metal goods	plastics	other	TOTAL
BASE	30	17	17	17	9	22	110
Before feasibility study	83	77	65	59	78	73	75
During feasibility study	10	18	18	41	22	23	19
After feasibility study	0	0	12	0	0	0	2
Not answered	7	6	6	0	0	5	5
TOTAL	100	100	100	100	100	100	100

Q.37 × Q.3

Table 44 Source of awareness of support to help use robots of plants receiving grants by employment size of plant

column percentages

Source of first information on possibility of grant	1–99	100–199	200–499	500–999	1000–	TOTAL
BASE	*20*	*13*	*35*	*13*	*28*	*110*
Trade, technical press	50	69	57	46	50	54
Suppliers	25	39	14	23	29	24
Exhibitions, trade fairs, conferences, courses	10	0	20	8	14	13
Other companies in group	5	0	14	15	11	10
General press	5	0	20	8	4	10
Consultants	5	8	6	15	7	7
Radio, television	5	0	3	15	4	5
Other	5	0	3	15	11	7
Not answered	5	0	6	0	7	5
TOTAL	115*	116*	143*	135*	137*	135*

* Totals add to more than 100 because some plants gave more than one source.

Q.37 × Q.6

Table 45 Source of awareness of support to help use robots of plants receiving grants by industry group

column percentages

Source of first information on possibility of grant	mech. eng.	elec. eng.	vehicles aircraft ships	other metal goods	plastics	other	TOTAL
BASE	*30*	*17*	*17*	*17*	*9*	*22*	*110*
Trade, technical press	57	35	47	47	67	59	54
Suppliers	23	29	35	29	11	23	24
Exhibitions, trade fairs, conferences, courses	10	24	12	12	33	5	13
Other companies in group	13	6	6	6	22	9	10
General press	20	18	6	0	0	0	10
Consultants	10	18	6	0	11	0	7
Radio, television	7	12	6	0	0	0	5
Other	3	24	0	12	0	0	7
Not answered	0	0	12	12	0	5	5
TOTAL	**143***	**166***	**130***	**118***	**144***	**101***	**135***

* Totals add to more than 100 because some respondents gave more than one source.

Q.36 × Q.3

Table 46 Reasons of non-applicants for not applying for government support to help with the purchase and development costs of robots by employment size of plant

column percentages

Reasons for not applying	1–99	100–199	200–499	500–999	1000–	TOTAL
BASE	*11*	*14*	*33*	*29*	*50*	*138*
Not aware support available	9	14	15	14	14	14
Not eligible for support	9	21	36	21	28	26
Did not need support	9	0	9	14	12	10
Level of support too low	0	7	0	7	4	4
Conditions for support too stringent	45	0	9	7	8	10
Processing of applications too slow	36	0	9	7	14	12
Other	0	43	9	17	10	14
Not answered	18	14	15	17	16	17
TOTAL	126*	100	102*	104*	106*	107*

* *Some totals add to more than 100 because some respondents gave more than one reason for not applying for support.*

Q.36 × Q.6

Table 47 Reasons of non-applicants for not applying for government support to help with the purchase and development costs of robots by industry group

column percentages

Reasons for not applying	mech. eng.	elec. eng.	vehicles aircraft ships	other metal goods	plastics	other	TOTAL
BASE	*27*	*19*	*38*	*14*	*14*	*31*	*138*
Not aware support available	11	11	11	14	29	16	14
Not eligible for support	26	21	45	14	14	16	26
Did not need support	0	21	8	14	7	13	10
Level of support too low	0	11	3	7	0	3	4
Conditions for support too stringent	19	11	11	7	0	7	10
Processing of applications too slow	26	11	13	14	0	3	12
Other	11	16	5	7	14	26	14
Not answered	22	5	13	21	36	16	17
TOTAL	115*	107*	109*	100	100	100	107*

* Some totals add to more than 100 because some respondents gave more than one reason for not applying for support.

Q.38 × Q.3

Table 48 Course firms would have taken in the absence of support for use of robots by employment size of plant

column percentages

Course firms would have taken in absence of grants	1–99	100–199	200–499	500–999	1000–	TOTAL
BASE	20	13	35	13	28	110
Not gone ahead at all	50	62	43	69	25	45
Gone ahead, but later	25	31	31	15	39	31
Gone ahead, but on a smaller scale	20	8	31	15	36	25
Not answered	10	0	0	0	11	5
TOTAL	105*	100	105*	100	111*	106*

* Some totals add to more than 100 because some respondents said their robots project would have been both smaller and later in the absence of support.

Q.38 × Q.6

Table 49 Course firms would have taken in the absence of support for use of robots by industry group

column percentages

Course firms would have taken in absence of grants	mech. eng.	elec. eng.	vehicles aircraft ships	other metal goods	plastics	other	TOTAL
BASE	30	17	17	17	9	22	110
Not gone ahead at all	67	41	53	35	44	55	45
Gone ahead, but later	27	29	35	41	22	18	31
Gone ahead, but on a smaller scale	13	18	29	24	22	18	25
Not answered	0	12	0	6	11	9	5
TOTAL	107*	100	117*	106*	100	100	106*

* Some totals add to more than 100 because some respondents said their project would have been both smaller and later in the absence of support.

Q.14

Table 50 Year first robot acquired by whether received support for development of robot application

cumulative column percentages, excluding not answered

Year	plants with grants to assist robot applications	other plants	TOTAL
BASE	*103*	*136*	*239*
Before 1975	1	4	2
1975–1979	8	13	11
1980	8	18	14
1981	9	28	20
1982	18	39	30
1983	40	52	47
1984	80	73	76
1985	100	100	100

Note: Government grants for robotics were introduced in 1981.

Q.39 × Q.3

Table 51 Willingness of grant recipients to undertake a further similar project without a grant by employment size of plant

column percentages

	1–99	100–199	200–499	500–999	1000–	TOTAL
BASE	*20*	*13*	*35*	*13*	*28*	*110*
Willingness to undertake a further similar project without a grant						
Yes	50	46	49	54	64	54
No	40	54	49	46	32	43
Not answered	10	0	3	0	4	4
TOTAL	100	100	100	100	100	100

Q.39 × Q.6

Table 52 Willingness of grant recipients to undertake a further similar project without grant by industry group

column percentages

	mech. eng.	elec. eng.	vehicles aircraft ships	other metal goods	plastics	other	TOTAL
BASE	*30*	*17*	*17*	*17*	*9*	*22*	*110*
Willingness to undertake a further similar project without a grant							
Yes	50	47	65	53	44	50	54
No	50	53	35	35	56	41	43
Not answered	0	0	0	12	0	9	4
TOTAL	100	100	100	100	100	100	100

Q.13

Table 53 Stage of use of first robot by whether received support for development of robot application

column percentages

	plants with grants to assist robot applications	other plants	TOTAL
BASE	*109*	*139*	*248*
Experimental, under development, pre-production	16	22	19
Installed for commercial production	78	70	73
Abandoned, permanently out of use or sold	5	8	7
Not answered	2	0	1
TOTAL	100	100	100

Q.17

Table 54 Number of robots per plant by whether received support for development of robot application

Number of robots per plant	firms with grant	firms without grant	TOTAL	Total number of users in UK
BASE	135	191	326	
1	43	43	43	320
2	20	17	18	130
3	10	8	9	70
4	3	8	6	40
5	5	4	4	30
6–10	9	7	8	60
11–20	3	7	5	40
21–30	2	2	2	20
over 30	0	1	1	10
Not answered	5	3	4	
TOTAL	100	100	100	740

Note: Figures for the total number of robot users in UK should be regarded as broadly indicative only. Particulars of the assumptions on which they are based are given in Appendix I.

NQ.10 × Q.8

Table 55 Non-users' awareness and use of grants for feasibility studies by plans to use robots

column percentages

Awareness and use of grant	PLANS TO USE ROBOTS			
	Yes in next two years	under consideration	No	TOTAL
BASE	*54*	*187*	*120*	*363*
Not aware grant available	17	31	35	30
Aware, but have not applied	43	57	63	57
Intend to apply for grant in future	19	7	1	7
Applied for grant, but application not yet decided on	4	0	0	1
Applied for grant, but application not accepted	4	1	0	1
Received grant for feasibility study	13	4	1	4
Not answered	2	1	1	1
TOTAL	**100**	**100**	**100**	**100**

NQ.8 × NQ.10

Table 56 Non-users' plans to use robots in next two years by awareness of grants for robots

column percentages

Plans to use robots	AWARENESS OF AVAILABILITY OF GRANTS				
	feasibility studies	robot costs	associated equipment	development work	TOTAL
BASE	*252*	*213*	*89*	*106*	*461*
Yes, in next two years	17	16	19	19	15
Under consideration	51	54	54	58	49
No	30	29	27	23	35
Not answered	1	1	0	0	1
TOTAL	100	100	100	100	100

Note: Grants for feasibility/planning studies are still available, but grants towards the cost of installation have ceased.

NQ.11 × Q.3

Table 57 Non-users' awareness of grants for investment in robots by employment size of plant

column percentages

Awareness of grants	1–99	100–199	200–499	500–999	1000–	TOTAL
BASE	*114*	*57*	*88*	*62*	*41*	*363*
Grants towards cost of robot	54	53	65	68	51	59
Grants towards cost of associated equipment	20	25	25	27	29	25
Grants for development work to get robot into production	19	32	31	32	44	29

Q.17

Table 58 Number of robots per plant

column percentages of plants, mean numbers of robots per plant, approximate total numbers of robots in UK and percentages of total numbers of robots in UK

Number of robots per plant	PLANTS		ROBOTS		
	per cent of plants in sample	total number of plants in UK	mean number of robots per plant	total number of robots in UK	per cent of total robots in UK
	%	no.	no.	no.	%
BASE	326				3208
1	43	320	1	350	10
2	18	130	2	300	9
3	9	70	3	200	6
4	6	40	4	200	6
5	4	30	5	150	5
6–10	8	60	8	500	16
11–20	5	40	15	600	18
Over 20	3	20	44	900	28
Not answered	4	30			
TOTAL	100	740	4.3	3208	100

Note: Figures for the total number of robots in UK should be regarded as broadly indicative only. Particulars of the assumptions on which they are based are given in Appendix I.

Q.17 × Q.14

Table 59 Number of robots per plant and total number of robots by year first robot acquired

column percentages

Number of robots per plant	before 1981	1981–1982	1983	1984	1985	TOTAL
BASE	33	38	41	70	57	326
1	12	21	34	47	70	43
2	15	18	10	24	12	18
3–5	27	16	34	16	11	19
6–10	15	13	12	6	2	8
Over 10	24	26	10	0	0	8
Not answered	6	5	0	7	5	4
TOTAL	100	100	100	100	100	100
Mean number of robots per plant	8	7	7	2	2	4.3
Total number of robots in UK	800	750	900	450	300	3208

Note: Figures for the total number of robots in UK should be regarded as broadly indicative only. Particulars of the assumptions on which they are based are given in Appendix I.

Q.17 × Q.1

Table 60 Number of robots per plant and total number of robots by type of company

column percentages

Number of robots per plant	UK company	overseas company	TOTAL
BASE	197	47	326
1	42	43	43
2	20	4	18
3–5	18	19	19
6–10	7	13	8
Over 10	9	13	8
Not answered	4	9	4
TOTAL	100	100	100
Mean number of robots per plant	4	7	4.3
Total number of robots in UK	2250	950	3208

Note: Figures for the total number of robots in UK should be regarded as broadly indicative only. Particulars of the assumptions on which they are based are given in Appendix I.

Q.17 × Q.2

Table 61 Number of robots per plant and total number of robots by number of plants of company

column percentages

Number of robots per plant	NUMBER OF UK PLANTS				
	1	2	3–5	5–	TOTAL
BASE	75	39	46	85	326
1	53	49	39	32	43
2	19	13	24	13	18
3–5	13	15	17	26	19
6–10	5	10	9	8	8
Over 10	5	10	7	14	8
Not answered	4	3	4	7	4
TOTAL	100	100	100	100	100
Mean number of robots per plant	3	4	4	7	4.3
Total number of robots in UK	600	450	500	1600	3208

Note: Figures for the total number of robots in UK should be regarded as broadly indicative only. Particulars of the assumptions on which they are based are given in Appendix I.

Q.17 × Q.3

Table 62 Number of robots per plant and total number of robots by employment size of plant

column percentages

Number of robots per plant	1–99	100–199	200–499	500–999	1000–	TOTAL
BASE	40	39	91	50	104	326
1	65	67	43	36	30	43
2	20	13	17	24	18	18
3–5	10	15	19	18	24	19
6–10	0	3	9	14	9	8
Over 10	0	0	11	4	14	8
Not answered	5	3	2	4	6	4
TOTAL	100	100	100	100	100	100
Mean number of robots per plant	2	2	4	4	7	4.3
Total number of robots in UK	150	150	800	400	1700	3208

Note: Figures for the total number of robots in UK should be regarded as broadly indicative only. Particulars of the assumptions on which they are based are given in Appendix I.

Q.17 × Q.6

Table 63 Number of robots per plant by industry group

column percentages

Number of robots per plant	mech. eng.	elec. eng.	vehicles aircraft ships	other metal goods	plastics	other	TOTAL
BASE	81	52	73	35	30	65	326
1	51	46	29	46	30	48	43
2	20	17	15	20	17	15	18
3–5	10	25	25	20	23	20	19
6–10	6	4	11	9	10	8	8
Over 10	4	2	16	3	20	5	8
Not answered	7	6	4	3	0	5	4
TOTAL	100	100	100	100	100	100	100
Mean number of robots per plant	3	3	7	3	7	4	4.3
Total number of robots in UK	450	300	1350	200	450	450	3208

Note: Figures for the total number of robots in UK should be regarded as broadly indicative only. Particulars of the assumptions on which they are based are given in Appendix I.

Q.17 × Q.4

Table 64 Number of robots per plant and total number of robots by number of shifts a day

column percentages

	NUMBER OF SHIFTS			
Number of robots per plant	one	two	three	TOTAL
BASE	81	100	55	326
1	57	39	27	43
2	21	17	11	18
3–5	14	16	27	19
6–10	5	9	11	8
Over 10	1	12	18	8
Not answered	3	7	6	4
TOTAL	100	100	100	100
Mean number of robots per plant	2	6	6	4.3
Total number of robots in UK	500	1650	1000	3208

Note: Figures for the total number of robots in UK should be regarded as broadly indicative only. Particulars of the assumptions on which they are based are given in Appendix I.

Q.20 × Q.4

Table 65 Hours a week robot scheduled to be running by number of shifts a day

column percentages

Hours a week	NUMBER OF SHIFTS			TOTAL
	one	two	three	
BASE	*81*	*100*	*55*	*248*
1–10	4	1	0	2
11–20	10	2	0	4
21–30	6	5	4	5
31–40	37	6	7	17
41–50	10	5	2	6
51–60	4	12	2	7
61–70	3	4	0	2
71–100	3	52	31	29
101–168	3	0	44	11
Not answered	22	13	11	17
TOTAL	100	100	100	100
Mean number of hours	42	69	100	68

Q.20,4 × Q.3

Table 66 Hours a week robot scheduled to be running and number of shifts worked by employment size of plant

column percentages

	1–99	100–199	200–499	500–999	1000–	TOTAL
BASE	31	27	68	42	78	248
One shift	65	48	28	26	23	33
Two shifts	26	22	44	41	50	40
Three shifts	10	22	27	26	22	22
Mean hours a week	50	65	73	71	72	68

Q.20,4 × Q.6

Table 67 Hours a week robot scheduled to be running and number of shifts worked by industry group

column percentages

	mech. eng.	elec. eng.	vehicles aircraft ships	other metal goods	plastics	other	TOTAL
BASE	57	36	55	31	23	53	248
One shift	44	42	22	32	9	32	33
Two shifts	46	31	56	55	17	34	40
Three shifts	9	25	16	7	70	28	22
Mean hours a week	57	68	67	63	100	70	68

Q.19

Table 68 Robot applications

Robot application	Plants with application		Robot application	plants with application
	per cent	total no. in UK		per cent
BASE	326		BASE	326
Arc welding	32	240	Casting	3
Assembly	18	130	Drilling	2
Machine loading	16	120	Cleaning	2
Painting, coating	14	100	Forging	1
Handling	14	100	Heat treatment	1
Injection moulding	8	60	Water cutting	1
Glueing, sealing	7	40	Laser cutting	1
Spot welding	6	40	Woodworking	1
Inspection	6	40	Food processing	1
Press loading	5	30	Textile handling	1
Grinding	5	30	Other	10
Training	5	30		

Note: Many users gave more than one robot application, making a total of 525 applications for the 326 plants in the sample, an average of 1.87 applications per plant.

Note: Figures for the total number of robot users in UK should be regarded as broadly indicative only. Particulars of the assumptions on which they are based are given in Appendix I.

Q.19 × Q.3

Table 69 Robot applications by employment size of plant

column percentages

	1–99	100–199	200–499	500–999	1000–	TOTAL
BASE	*40*	*39*	*91*	*50*	*104*	*326*
Arc welding	48	41	36	16	26	32
Spot welding	0	0	9	0	13	6
Painting, coating	3	10	15	16	18	14
Glueing, sealing	3	5	4	4	12	7
Assembly	15	8	13	18	27	18
Handling, packaging	5	10	14	8	21	17
Grinding	3	5	6	4	6	5
Injection moulding	5	15	13	6	4	8
Press loading	8	8	6	8	2	5
Machine loading	5	8	12	18	28	16

Q.3 × Q.19

Table 70 Robot applications distribution between plants of different employment sizes

row percentages

Robot application	BASE	1–99	100–199	200–499	500–999	1000–	TOTAL
Arc welding	*105*	18	15	31	8	26	100
Spot welding	*21*	0	0	38	0	62	100
Painting, coating	*46*	2	9	30	17	41	100
Glueing, sealing	*22*	5	9	18	9	54	100
Assembly	*58*	10	5	21	16	48	100
Handling, packaging	*51*	4	8	28	14	47	100
Grinding	*16*	6	13	31	13	38	100
Injection moulding	*27*	7	22	44	11	15	100
Press loading	*17*	18	18	29	24	12	100
Machine loading	*51*	4	6	22	18	51	100

Q.19 × Q.6

Table 71 Robot applications by industry group

column percentages

Robot application	mech. eng.	elec. eng.	vehicles aircraft ships	other metal goods	plastics	other	TOTAL
BASE	*81*	*52*	*73*	*35*	*30*	*65*	*326*
Arc welding	51	23	40	54	0	12	32
Spot welding	6	4	16	9	0	2	6
Painting, coating	10	12	26	17	23	6	14
Glueing, sealing	0	12	11	0	3	11	7
Assembly	11	37	22	6	17	14	18
Handling, packaging	9	12	15	6	17	37	17
Grinding	5	2	8	6	3	3	5
Injection moulding	1	4	6	3	63	5	8
Press loading	7	4	3	9	7	3	5
Machine loading	22	12	26	17	7	8	16

Q.6 × Q.19

Table 72 Robot applications' distribution between industry groups

row percentages

Robot application	mech. eng.	elec. eng.	vehicles aircraft ships	other metal goods	plastics	other	TOTAL
BASE							
Arc welding	105	39	11	28	18	0	8	100
Spot welding	21	24	10	57	14	0	5	100
Painting, coating	46	17	13	41	13	15	9	100
Glueing, sealing	22	0	27	36	0	5	32	100
Assembly	58	16	33	28	3	9	16	100
Handling, packaging	51	14	10	22	4	8	43	100
Grinding	16	25	6	38	13	6	13	100
Injection moulding	27	4	7	15	4	70	11	100
Press loading	17	35	12	12	18	12	12	100
Machine loading	51	35	12	37	12	4	10	100

Q.19 × Q.17

Table 73 Robot applications by total number of robots in use in plant

column percentages of plants and mean numbers of robots per plant

	robots per plant			TOTAL	mean number of robots per plant
	1	2–5	6+		
	%	%	%	%	no.
BASE	140	121	52	326	
Arc welding	35	24	40	32	5
Spot welding	1	7	17	6	22
Painting, coating	6	20	21	14	9
Glueing, sealing	5	3	17	7	21
Assembly	12	17	31	18	9
Handling, packaging	10	17	35	17	7
Grinding	4	4	8	5	5
Injection moulding	4	10	17	8	8
Press loading	4	7	8	5	4
Machine loading	8	21	23	16	5

Note: Figures are for all kinds of robots in the plant including those working on other applications.

Q.17 × Q.19

Table 74 Robot applications' distribution between plants with different numbers of robots

row percentages

Robot application		Robots per Plant					TOTAL	mean number per plant
		1	2	3–5	6–10	over 10		
	BASE							
Arc welding	105	47	13	14	9	11	100	5
Spot welding	21	5	14	29	10	33	100	22
Painting, coating	46	20	30	22	9	15	100	9
Glueing, sealing	22	32	5	14	5	36	100	21
Assembly	58	29	14	22	10	17	100	9
Handling, packaging	51	26	14	24	14	18	100	7
Grinding	16	38	0	31	13	13	100	5
Injection moulding	27	19	7	37	11	22	100	8
Press loading	17	29	24	24	18	6	100	4
Machine loading	51	22	20	31	14	10	100	5

Note: Figures are for all kinds of robots in the plant, including those working on other applications.

NQ.12 x NQ.8

Table 75 Robot applications envisaged by non-users

Robot application envisaged	PLANS TO USE ROBOTS IN FUTURE			
	Yes, in next 2 years	under consideration	No	TOTAL
BASE	*68*	*226*	*161*	*461*
Assembly	38	32	14	26
Machine loading	28	24	12	20
Handling	15	27	13	20
Arc welding	24	15	5	13
Packaging	9	13	8	10
Press loading	9	11	6	9
Painting, coating	9	10	4	7
Inspection	3	10	6	7
Injection moulding	12	7	4	7
Grinding	10	4	3	4
Glueing, sealing	4	6	2	4
Drilling	2	5	4	4
Spot welding	2	6	1	4
Training	3	3	2	3
Cleaning	1	3	2	2
Laser cutting	2	2	3	2
Casting	0	3	1	2
Food processing	0	2	1	1
Forging	0	1	1	1
Textile handling	0	1	1	1
Woodworking	0	1	0	1
Heat treatment	0	1	1	1

Q.27 × Q.13

Table 76 Attitude of the workers directly affected by the introduction of robots by stage of use of first robot

column percentages

Attitudes of workers affected	experimental, pre-production	installed for commercial production	abandoned or sold	TOTAL
BASE	48	182	16	248
BEFORE INTRODUCTION OF ROBOTS				
Very favourable	19	13	13	14
Quite favourable	31	27	31	28
Neutral	29	46	25	41
Quite unfavourable	4	9	19	9
Very unfavourable	0	1	0	1
Not answered	17	4	13	7
TOTAL	100	100	100	100
AFTER INTRODUCTION OF ROBOTS				
Very favourable	21	35	25	32
Quite favourable	15	47	38	40
Neutral	13	14	19	14
Quite unfavourable	2	2	6	2
Very unfavourable	0	1	6	1
Not answered	50	2	6	11
TOTAL	100	100	100	100

Q.27 × Q.17

Table 77 Attitude of the workers directly affected by the introduction of robots by number of robots in use

column percentages of plants, mean numbers of robots per plant, approximate total numbers of robots in UK and percentages of total numbers of robots in UK

Attitude of workers directly affected	PLANTS				ROBOTS		
	robots per plant				mean number of robots per plant	total number of robots in UK	per cent of total robots in UK
	1	2–5	6+	TOTAL			
	%	%	%	%	no.	no.	%
BASE	104	89	43	248			3208
BEFORE INTRODUCTION OF ROBOTS							
Favourable	44	42	42	42	5	1450	46
Neutral	40	40	44	41	5	1300	40
Unfavourable	7	9	14	9	5	350	11
Not answered	9	9	0	7	2	100	3
TOTAL	100	100	100	100	4.3	3208	100
AFTER INTRODUCTION OF ROBOTS							
Favourable	65	73	84	71	5	2500	78
Neutral	15	12	14	14	7	500	15
Unfavourable	6	2	0	4	1	50	1
Not answered	14	12	2	11	2	150	5
TOTAL	100	100	100	100	4.3	3208	100

Note: Figures for the total number of robots in UK should be regarded as broadly indicative only. Particulars of the assumptions on which they are based are given in Appendix I.

Q.27 x Q.28,29

Table 78 Attitude of the workers directly affected by the introduction of robots by whether there was advance consultation and whether there were negotiations with the unions

column percentages

Attitude of workers directly affected	Advance Consultation		Union Negotiations		TOTAL
	yes	no	yes	no	
BASE	*194*	*44*	*60*	*174*	*248*
BEFORE INTRODUCTION OF ROBOTS					
Favourable	47	27	37	45	42
Neutral	41	52	43	43	41
Unfavourable	8	16	15	8	9
Not answered	5	5	5	5	7
TOTAL	100	100	100	100	100
AFTER INTRODUCTION OF ROBOTS					
Favourable	78	54	63	77	71
Neutral	11	30	25	10	14
Unfavourable	3	7	3	3	4
Not answered	8	9	8	9	11
TOTAL	100	100	100	100	100

Q.27 × Q.30

Table 79 Attitude of the workers directly affected by the introduction of robots by changes in employment as a direct result of the introduction of robots

column percentages

Attitude of workers directly affected	increase in jobs	no change in jobs	decrease in jobs	TOTAL
BASE	*20*	*148*	*62*	*248*
BEFORE INTRODUCTION OF ROBOTS				
Favourable	60	45	37	42
Neutral	30	45	39	41
Unfavourable	10	7	18	9
Not answered	0	3	7	7
TOTAL	100	100	100	100
AFTER INTRODUCTION OF ROBOTS				
Favourable	90	72	77	71
Neutral	0	16	15	14
Unfavourable	5	4	2	4
Not answered	5	8	7	11
TOTAL	100	100	100	100

Q.28,29 × Q.13

Table 80 Consultation with those directly affected and negotiation with trade unions by stage of use of first robot

column percentages

	experimental, pre-production	installed for commercial production	abandoned or sold	TOTAL
BASE	48	182	16	248
Consultation with those directly affected by introduction of robots				
Yes	69	82	63	78
No	17	17	31	18
Not answered	15	1	6	4
TOTAL	100	100	100	100
Negotiations with trade unions on introduction of robots				
Yes	17	26	25	24
No	67	73	56	70
Not answered	17	1	19	6
TOTAL	100	100	100	100

Q.28,29 × Q.30

Table 81 Consultation with those directly affected and negotiation with trade unions by changes in employment as a direct result of the introduction of robots

column percentages

	increase in jobs	no change in jobs	decrease in jobs	TOTAL
BASE	*20*	*148*	*62*	*248*
Consultation with those directly affected by the introduction of robots				
Yes	85	79	84	78
No	10	20	16	18
Not answered	5	1	0	4
TOTAL	100	100	100	100
Negotiations with trade unions on introduction of robots				
Yes	0	25	32	24
No	95	72	66	70
Not answered	5	3	2	6
TOTAL	100	100	100	100

Q.30 × Q.13

Table 82 Changes in employment as a direct result of the introduction of robots by stage of use of first robot

column percentages of plants and mean and total number of jobs

	experimental, pre-production	installed for commercial production	abandoned or sold	TOTAL
BASE	48	182	16	248
Proportion of plants with				
Increase in jobs	4	10	0	8
No change in jobs	56	61	56	60
Decrease in jobs	17	27	25	25
Not answered	23	2	19	7
TOTAL	100	100	100	100
Change in number of jobs				
Mean increase per plant in plants with increase				+13
Mean decrease per plant in plants with decrease				−8
Mean change in all plants				−1
Total UK increase in plants with increase				+700
Total UK decrease in plants with decrease				−1400
Total UK change in all plants using robots				−700

Q.30 × Q.1

Table 83 Changes in employment as a direct result of the introduction of robots by type of company

column percentages of plants and mean numbers of jobs

	UK company	overseas company	TOTAL
BASE	197	47	248
Proportion of plants with			
Increase in jobs	9	6	8
Decrease in jobs	23	32	25
Mean change in number of jobs per plant	−1	−2	−1

Q.30 × Q.3

Table 84 Changes in employment as a direct result of the introduction of robots by employment size of plant

column percentages of plants and mean numbers of jobs

	1–99	100–199	200–499	500–999	1000–	TOTAL
BASE	31	27	68	42	78	248
Percentage of plants with						
Increase in jobs	26	11	6	5	4	8
Decrease in jobs	10	26	22	38	27	25
Mean change in number of jobs per plant	+1	–1	0	–3	–3	–1

Q.30 × Q.6

Table 85 Changes in employment as a direct result of the introduction of robots by industry group

column percentages of plants and mean numbers of jobs

	mech. eng.	elec. eng.	vehicles aircraft ships	other metal goods	plastics	other	TOTAL
BASE	57	36	55	31	23	53	248
Percentage of plants with							
Increase in jobs	11	6	4	23	4	9	8
Decrease in jobs	19	36	24	32	26	17	25
Mean change in number of jobs per plant	–1	–4	–2	+1	–1	–1	–1

Q.30 × Q.14

Table 86 Changes in employment as a direct result of the introduction of robots by year first robot acquired

column percentages

	before 1981	1981–1982	1983	1984	1985	TOTAL
BASE	33	38	41	70	57	248
Percentage of plants with						
Increase in jobs	3	11	2	10	12	8
Decrease in jobs	42	42	24	17	16	25
Mean change in number of jobs per plant	−6	0	−2	0	0	−1

Q.30 × Q.17

Table 87 Changes in employment as a direct result of the introduction of robots by number of robots in use

column percentages of plants, mean numbers of robots per plant, approximate total numbers of robots in UK and percentages of total numbers of robots in UK

	PLANTS					ROBOTS	
	robots per plant				mean number of robots per plant	total number of robots in UK	per cent of total robots in UK
	1	2–5	6+	TOTAL			
	%	%	%	%	no.	no.	%
BASE	104	89	43	248			3208
Percentage of plants with							
Increase in jobs	9	6	9	8	4	250	8
Decrease in jobs	14	31	40	25	5	900	28
Mean change in number of jobs per plant	0	−2	−3	−1			

Q.31

Table 88 Numbers already been on robot training courses and numbers planned to go in next two years by type of organisation providing training course

column percentages of robot user plants and mean and total numbers on courses

		own staff in-house	other company in group	robot supplier	specialist training organisation	other	all
BASE		248	248	248	248	248	248
ON COURSE ALREADY							
One or more been on course	%	27	5	76	10	2	
None been on course	%	73	95	24	90	98	
Mean number on course:							
plants using courses	no.	9	14	6	8	61	
all robot users	no.	2	1	5	1	1	10
Total number in UK been on course (approx.)	no.	1800	500	3500	600	700	7100
ON COURSE IN NEXT TWO YEARS							
One or more on course	%	21	2	33	5	1	
None to go on course	%	79	98	67	95	99	
Mean number on course:							
plants using courses	no.	10	20	6	6	36	
all robot users	no.	2	1	2	0	0	5
Total number in UK to go on course (approx.)	no.	1400	400	1600	300	300	4000

Q.31 × Q.3

Table 89 Numbers already been on robot training courses and numbers planned to go in next two years by employment size of plant

mean and total numbers on courses

	1–99	100–199	200–499	500–999	1000–	TOTAL
BASE	31	27	68	42	78	248
ON COURSE ALREADY						
Mean number per plant, all robot users	4	5	10	7	14	10
Total number in UK (approx.)	400	400	2100	900	3300	7100
ON COURSE IN NEXT TWO YEARS						
Mean number per plant, all robot users	1	2	6	4	9	5
Total number in UK (approx.)	100	200	1200	500	2000	4000

Q.31 × Q.6

Table 90 Numbers already been on robot training courses and numbers planned to go in next two years by industry group

mean and total numbers on courses

	mech. eng.	elec. eng.	vehicles aircraft ships	other metal goods	plastics	other	TOTAL
BASE	57	36	55	31	23	53	248
ON COURSE ALREADY							
Mean number per plant, all robot users	7	5	17	9	8	9	10
Total number in UK (approx)	1100	500	2700	800	600	1400	7100
ON COURSE IN NEXT TWO YEARS							
Mean number per plant, all robot users	4	3	8	8	4	6	5
Total number in UK (approx)	600	300	1200	700	300	900	4000

Q.31 × Q.17

Table 91 Numbers already been on training courses and numbers planned to go in next two years by number of robots in use

mean and total numbers on courses

	1	2	3–5	6–10	Over 10	TOTAL
BASE	104	42	47	20	23	248
ON COURSE ALREADY						
Mean number per plant, all robot users	6	6	11	15	21	10
Total number in UK (approx)	1900	900	1700	1000	1600	7100
ON COURSE IN NEXT TWO YEARS						
Mean number per plant, all robot users	2	2	7	14	13	5
Total number in UK (approx)	800	300	1000	900	1000	4000

Q.32 × Q.3

Table 92 Extent of satisfaction with training arrangements by employment size of plant

column percentages

	1–99	100–199	200–499	500–999	1000–	TOTAL
BASE	31	27	68	42	78	248
Satisfied with training	48	37	47	43	32	40
Not satisfied with training	16	11	4	14	17	13
Other comments	10	7	12	5	15	11
Not answered	26	44	37	38	36	36
TOTAL	100	100	100	100	100	100

Q.32 × Q.6

Table 93 Extent of satisfaction with training arrangements by industry group

column percentages

	mech. eng.	elec. eng.	vehicles aircraft ships	other metal goods	plastics	other	TOTAL
BASE	57	36	55	31	23	53	248
Satisfied with training	46	39	36	39	48	36	40
Not satisfied with training	16	8	13	16	9	10	13
Other comments	5	8	16	13	13	9	11
Not answered	33	44	35	32	30	45	36
TOTAL	100	100	100	100	100	100	100

Q.24

Table 94 Difficulties and disadvantages expected by robot users before going into production compared with those actually experienced after

column percentages and rank orders

DIFFICULTIES EXPECTED				DIFFICULTIES EXPERIENCED	
Type of difficulty	per cent of users	rank order	rank order	per cent of users	Type of difficulty
BASE	248			248	BASE
Installation, integration	43	1	1	37	High costs of development
Opposition from shopfloor, unions	31	2	2	33	Inadequate after-sales support
High costs of development	26	3=	3	32	Installation, integration
High costs of equipment	26	3=	4	27	Lack of technical expertise
Lack of technical expertise	25	5	5	26	Reliability, maintenance
Reliability, maintenance	21	6	6	23	High costs of equipment
Inadequate after-sales support	8	7=	7	13	Unsuitability for task
Unsuitability for task	8	7=	8	9	No advantage over hard automation
No advantage over hard automation	6	9=	9=	6	Opposition from other groups
Health and safety	6	9=	9=	6	Health and safety
Opposition from other groups	4	11	11	4	No advantage over existing equipment
No advantage over existing equipment	4	12	12=	2	Opposition from top management
Opposition from top management	2	13	12=	2	Opposition from shopfloor, unions
Other problems	1			4	Other problems
No problems stated	0			26	No problems stated
TOTAL	212*			196*	TOTAL (excl. no problems)

* Totals add to much more than 100 because many users indicated more than one kind of difficulty.

Q.24 × Q.13

Table 95 Main difficulties and disadvantages expected by robot users compared with those actually experienced by stage of use of first robot

column percentages

Type of difficulty expected Type of difficulty experienced		(a) (b)	experimental, pre- production	installed for commercial production	abandoned or sold	TOTAL
BASE			48	182	16	248
High costs of equipment		(a)	21	28	25	26
		(b)	8	26	38	23
High costs of development		(a)	38	24	19	26
		(b)	27	39	38	37
Installation, integration		(a)	40	44	31	43
		(b)	25	33	38	32
Reliability, maintenance		(a)	17	21	31	21
		(b)	13	28	50	26
Lack of technical expertise		(a)	23	26	13	25
		(b)	23	28	31	27
Opposition from shop floor, unions		(a)	15	36	19	31
		(b)	0	3	6	2

Q.24 × Q.3

Table 96 Main difficulties and disadvantages expected by robot users compared with those actually experienced by employment size of plant

column percentages

Type of difficulty expected Type of difficulty experienced	(a) (b)	1–99	100– 199	200– 499	500– 999	1000–	TOTAL
BASE		31	27	68	42	78	248
High costs of equipment	(a)	23	15	29	24	31	26
	(b)	19	11	24	26	27	23
High costs of development	(a)	23	11	27	21	36	26
	(b)	29	30	38	41	39	37
Installation, integration	(a)	42	30	38	48	47	43
	(b)	16	41	31	29	37	32
Reliability, maintenance	(a)	16	11	28	21	21	21
	(b)	26	7	28	29	30	26
Lack of technical expertise	(a)	23	22	29	19	26	25
	(b)	23	15	31	24	30	27
Opposition from shop floor, unions	(a)	19	22	32	26	40	31
	(b)	0	0	0	5	5	2

Q.24 × Q.6

Table 97 Main difficulties and disadvantages expected by robot users compared with those actually experienced by industry group

column percentages

Type of difficulty expected Type of difficulty experienced	(a) (b)	mech. eng.	elec. eng.	vehicles aircraft ships	other metal goods	plastics	other	TOTAL
BASE		57	36	55	31	23	53	248
High costs of equipment	(a)	28	25	31	42	13	25	26
	(b)	23	19	26	42	9	15	23
High costs of development	(a)	32	19	26	32	9	34	26
	(b)	44	44	29	39	22	34	37
Installation, integration	(a)	46	53	40	39	26	45	43
	(b)	39	36	35	26	26	28	32
Reliability, maintenance	(a)	16	14	29	16	22	25	21
	(b)	23	28	35	19	26	21	26
Lack of technical expertise	(a)	26	25	33	26	9	23	25
	(b)	30	19	40	32	22	17	27
Opposition from shop floor, unions	(a)	28	25	36	42	17	26	31
	(b)	2	3	4	0	0	4	2

Q.24 × Q.10

Table 98 Main difficulties and disadvantages expected by robot users compared with those actually experienced by type of feasibility study

column percentages

Type of difficulty expected Type of difficulty experienced	(a) (b)	in-house	company in group	robot supplier	consul- tant	none	TOTAL
BASE		93	18	75	35	24	248
High costs of equipment	(a)	25	17	28	29	29	26
	(b)	20	17	24	31	21	23
High costs of development	(a)	24	22	29	26	25	26
	(b)	26	28	43	54	42	37
Installation, integration	(a)	40	50	39	57	38	43
	(b)	36	22	24	43	25	32
Reliability, maintenance	(a)	20	28	24	20	17	21
	(b)	27	22	23	29	29	26
Lack of technical expertise	(a)	25	28	24	31	21	25
	(b)	19	33	29	34	29	27
Opposition from shop floor, unions	(a)	32	39	31	31	33	31
	(b)	3	6	4	0	0	2

Q.24 × Q.14

Table 99 Main difficulties and disadvantages expected by robot users compared with those actually experienced by year first robot acquired

column percentages

Type of difficulty expected Type of difficulty experienced	(a) (b)	before 1981	1981– 1982	1983	1984	1985	TOTAL
BASE		*33*	*38*	*41*	*70*	*57*	*248*
High costs of equipment	(a)	42	16	27	29	23	26
	(b)	36	26	24	26	11	23
High costs of development	(a)	24	16	29	31	26	26
	(b)	33	34	49	43	23	37
Installation, integration	(a)	42	42	37	41	49	43
	(b)	39	32	32	29	33	32
Reliability, maintenance	(a)	30	26	15	20	19	21
	(b)	36	24	32	34	5	26
Lack of technical expertise	(a)	24	18	20	29	28	25
	(b)	33	24	27	30	21	27
Opposition from shop floor, unions	(a)	52	16	22	34	32	31
	(b)	0	3	0	6	2	2

Q.24 × Q.13

Table 100 Difficulties and disadvantages experienced by robot users by stage of use of first robot

column percentages

Type of difficulty	experimental, pre-production	installed for commercial production	abandoned or sold	TOTAL
BASE	*48*	*182*	*16*	*248*
High costs of equipment	8	26	38	23
High costs of development	27	39	38	37
Installation, integration	25	33	38	32
Reliability, maintenance	13	28	50	26
Inadequate after-sales support	19	35	44	33
Lack of technical expertise	23	28	31	27
Unsuitability for task	10	12	25	13
No advantage over hard automation	6	8	25	9
No advantage over existing equipment	2	4	13	4
Health and safety	2	6	25	6
Opposition from shop floor, unions	0	3	6	2
Opposition from top management	2	3	0	2
Opposition from other groups	6	6	6	6
Other problems	0	5	0	4
No problems stated	50	20	25	26
TOTAL (excl. no problems)	143*	213*	313*	196*

* Totals add to more than 100 because many users indicated more than one kind of difficulty.

Q.24 × Q.3

Table 101 Difficulties and disadvantages experienced by robot users by employment size of plant

column percentages

Type of difficulty	1–99	100–199	200–499	500–999	1000–	TOTAL
BASE	31	27	68	42	78	248
High costs of equipment	19	11	24	26	27	23
High costs of development	29	30	38	41	39	37
Installation, integration	16	41	31	29	37	32
Reliability, maintenance	26	7	28	29	30	26
Inadequate after-sales support	19	19	44	31	32	33
Lack of technical expertise	23	15	31	24	30	27
Unsuitability for task	16	4	15	12	12	13
No advantage over hard automation	7	7	9	14	6	9
No advantage over existing equipment	4	0	7	5	3	4
Health and safety	3	4	7	2	9	6
Opposition from shop floor, unions	0	0	0	5	5	2
Opposition from top management	3	0	2	2	4	2
Oppostion from other groups	3	0	10	7	5	6
Other problems	3	15	2	2	3	4
No problems stated	32	30	25	29	23	26
TOTAL (excl. no problems)	139*	122*	219*	200*	217*	196*

* Total adds to more than 100 because many users indicated more than one kind of difficulty.

Q.24 × Q.6

Table 102 Difficulties and disadvantages experienced by robot users by industry group

column percentages

Type of difficulty	mech. eng.	elec. eng.	vehicles aircraft ships	other metal goods	plastics	other	TOTAL
BASE	57	36	55	31	23	53	248
High costs of equipment	23	19	26	42	9	15	23
High costs of development	44	44	29	39	22	34	37
Installation, integration	39	36	35	26	26	28	32
Reliability, maintenance	23	28	35	19	26	21	26
Inadequate after-sales support	37	36	35	42	26	13	33
Lack of technical expertise	30	19	40	32	22	17	27
Unsuitability for task	7	19	24	0	13	8	13
No advantage over hard automation	9	14	7	7	4	8	9
No advantage over existing equipment	4	8	4	3	0	4	4
Health and safety	2	0	9	10	17	4	6
Opposition from shop floor, unions	2	3	4	0	0	4	2
Opposition from top management	0	3	9	0	0	0	2
Opposition from other groups	4	11	7	7	0	6	6
Other problems	5	0	4	0	4	6	4
No problems stated	23	22	18	29	30	40	26
TOTAL (excl. no problems)	203*	220*	247*	197*	140*	126*	196*

* Totals add to more than 100 because many users indicated more than one kind of difficulty.

Q.24 × Q.10

Table 103   Difficulties and disadvantages experienced by robot users by type of feasibility study

column percentages

Type of difficulty	in-house	company in group	robot supplier	consul-tant	none	TOTAL
BASE	*93*	*18*	*75*	*35*	*24*	*248*
High costs of equipment	20	17	24	31	21	23
High costs of development	26	28	43	54	42	37
Installation, integration	36	22	24	43	25	32
Reliability, maintenance	27	22	23	29	29	26
Inadequate after-sales support	32	22	33	31	33	33
Lack of technical expertise	19	33	29	34	29	27
Unsuitability for task	13	0	13	14	8	13
No advantage over hard automation	5	17	8	11	8	9
No advantage over existing equipment	2	0	4	6	4	4
Health, and safety	3	6	7	11	8	6
Opposition from shop floor, unions	3	6	4	0	0	2
Opposition from top management	3	6	0	3	4	2
Opposition from other groups	9	6	3	9	4	6
Other problems	3	11	5	0	0	4
No problems stated	34	22	23	6	29	26
TOTAL (excl. no problems)	168*	172*	197*	271	188*	196*

*   Totals add to more than 100 because many users indicated more than one kind of difficulty.

Q.24 × Q.14

Table 104 Difficulties and disadvantages experienced by robot users by year first robot acquired

column percentages

Type of difficulty	before 1981	1981–1982	1983	1984	1985	TOTAL
BASE	33	38	41	70	57	248
High costs of equipment	36	26	24	26	11	23
High costs of development	33	34	49	43	23	37
Installation, integration	39	32	32	29	33	32
Reliability, maintenance	36	24	32	34	5	26
Inadequate after-sales support	42	34	37	40	12	33
Lack of technical expertise	33	24	27	30	21	27
Unsuitability for task	21	13	15	16	2	13
No advantage over hard automation	6	11	12	10	5	9
No advantage over existing equipment	3	5	2	7	2	4
Health and safety	12	5	2	10	2	6
Opposition from shop floor, unions	0	3	0	6	2	2
Opposition from top management	0	0	5	4	0	2
Opposition from other groups	6	5	7	7	5	6
Other problems	6	5	7	3	0	4
No problems stated	13	26	17	23	46	26
TOTAL (excl. no problems)	263*	195*	234*	241*	76*	196*

* Totals add to more than 100 because many users indicated more than one kind of difficulty.

Q.24 × Q.17

Table 105 Main difficulties and disadvantages experienced by robot users by number of robots in use

column percentages of plants, mean numbers of robots per plant, approximate total numbers of robots in UK and percentages of total numbers of robots in UK

	PLANTS				ROBOTS		
	robots per plant				mean number of robots per plant	total number of robots in UK	per cent of total robots in UK
	1	2–5	6+	TOTAL			
	%	%	%	%	no.	no.	%
BASE	*104*	*89*	*43*	*248*			*3208*
High costs of equipment	16	28	26	23	4	700	22
High costs of development	39	37	35	37	4	950	29
Installation, integration	34	24	42	32	5	1050	33
Reliability, maintenance	24	24	33	26	6	1050	32
Inadequate after-sales support	31	28	51	33	6	1350	41
Lack of technical expertise	20	28	33	27	6	950	29
Unsuitability for task	10	15	16	13	8	600	18
Opposition from shop floor, unions	2	3	0	2	2	50	1
No problems stated	31	26	16	26	4	850	27

NQ.14 × NQ.8

Table 106 Difficulties and disadvantages expected by non-users compared with difficulties and disadvantages expected and actually experienced by existing robot users

column percentages

	NON-USERS				USERS	
	yes, in next two years	under consideration	no	TOTAL	total (expected)	total (actual)
BASE	54	187	120	363	248	248
High costs of development	30	43	19	33	26	37
Inadequate after-sales support	19	9	8	10	8	33
Installation, integration	30	29	18	26	43	32
Lack of technical expertise	26	29	23	26	25	27
Reliability, maintenance	15	16	18	16	21	26
High costs of equipment	63	66	49	60	26	23
Unsuitability for task	24	40	40	38	8	13
No advantage over hard automation	11	19	16	17	6	9
Opposition from other groups	2	1	0	1	4	6
Health and safety	2	5	3	4	6	6
No advantage over existing equipment	11	12	21	15	4	4
Opposition from top management	4	7	3	5	2	2
Opposition from shop floor, unions	26	19	14	18	31	2
Other problems	2	4	3	3	1	4
No problems stated	6	5	19	10	0	26
TOTAL (excl. no problems)	265*	299*	235*	272*	212*	196*

* Totals add to much more than 100 because many users indicated more than one kind of difficulty.

Table 107 Disadvantages and problems experienced by robot users compared with users of other microelectronics-based production technologies

percentage of users* rating disadvantage very important

Disadvantage	all micro-electronics users	robot users all	1 robot only	2 or more robots
SAMPLE BASE	469	59	28	28
Lack of people with microelectronics expertise	49	61	52	72
High costs of development	30	26	37	15
Lack of finance for development	26	11	11	10
General economic situation	25	13	13	12
Problems with software	17	28	33	19
Problems with sensors	14	23	37	7
Problems with chips	14	15	17	9
Difficulties of communications with subcontractors or suppliers	12	21	29	13
Higher production costs	10	12	14	10
Opposition from shop floor or unions	4	2	0	5
Opposition in top management	3	4	3	5

* PSI survey of microelectronics applications in industry in 1985, sample of 1200 weighted to represent all factories in UK manufacturing employing 20 or more people.

Q.18 × Q.13

Table 108 Price of robots by stage of use

column percentages

Price of latest robot in £'000	experimental, pre-production	installed for commercial production	abandoned or sold	TOTAL
BASE	48	182	16	248
Under £10	10	7	13	8
£10–19	2	10	6	8
£20–29	29	12	13	15
£30–39	8	19	25	17
£40–49	8	14	6	13
£50 +	35	35	38	36
Not answered	6	3	0	3
TOTAL	100	100	100	100
Median price (£'000)	35	39	37	39

Q.18 × Q.6

Table 109 Price of robots by industry group

column percentages

Price of latest robot in £'000	mech. eng.	elec. eng.	vehicles aircraft ships	other metal goods	plastics	other	TOTAL
BASE	57	36	55	31	23	53	248
Under £10	5	11	4	3	30	11	8
£10–19	14	3	4	10	4	9	8
£20–29	14	19	11	10	9	25	15
£30–39	19	22	18	16	13	15	17
£40–49	7	11	16	3	17	17	13
£50 +	39	31	46	55	22	17	36
Not answered	2	3	2	3	4	6	3
TOTAL	100	100	100	100	100	100	100
Median price (£'000)	37	36	47	52	32	30	39

Q.18 × Q.10

Table 110 Price of robots by type of feasibility study

column percentages

Price of latest robot in £'000	own company in-house	other company in group	robot supplier	consul- tant	none	TOTAL
BASE	93	18	75	35	24	248
Under £10	9	11	4	11	8	8
£10–£19	10	6	9	3	8	8
£20–£29	19	0	13	14	25	15
£30–£39	16	11	16	26	17	17
£40–£49	11	17	15	14	4	13
£50–	32	44	40	31	33	36
Not answered	3	11	3	0	4	3
TOTAL	100	100	100	100	100	100
Median price (£'000)	36	49	43	37	33	39

Q.21 × Q.13

Table 111 Downtime of robots compared with expectations by stage of use of first robot

column percentages

Downtime	experimental, pre-production	installed for commercial production	abandoned or sold	TOTAL
BASE	48	182	16	248
More than expected	33	36	56	37
Same as expected	23	35	19	32
Less than expected	19	24	13	23
Not answered	25	4	13	9
TOTAL	100	100	100	100

Q.21 × Q.6

Table 112 Downtime of robots compared with expectations by industry group

column percentages

Downtime	mech. eng.	elec. eng.	vehicles aircraft ships	other metal goods	plastics	other	TOTAL
BASE	57	36	55	31	23	53	248
More than expected	42	44	38	29	30	32	37
Same as expected	32	25	33	39	26	28	32
Less than expected	21	25	29	19	26	23	23
Not answered	5	6	0	13	17	17	9
TOTAL	100	100	100	100	100	100	100

Q.21 × Q.10

Table 113   Downtime of robots compared with expectations by type of feasibility study

column percentages

Downtime	in-house	company in group	robot supplier	consul-tant	none	TOTAL
BASE	*93*	*18*	*75*	*35*	*24*	*248*
More than expected	**32**	**33**	**37**	**43**	**46**	**37**
Same as expected	**27**	**22**	**36**	**31**	**33**	**32**
Less than expected	**33**	**28**	**19**	**14**	**17**	**23**
Not answered	**8**	**17**	**8**	**11**	**4**	**9**
TOTAL	**100**	**100**	**100**	**100**	**100**	**100**

Q.21 × Q.14

Table 114   Downtime of robots compared with expectations by year first robot acquired

column percentages

Downtime	before 1981	1981–1982	1983	1984	1985	TOTAL
BASE	*33*	*38*	*41*	*70*	*57*	*248*
More than expected	**33**	**34**	**42**	**44**	**28**	**37**
Same as expected	**24**	**26**	**29**	**34**	**37**	**32**
Less than expected	**39**	**32**	**17**	**21**	**16**	**23**
Not answered	**3**	**8**	**12**	**0**	**19**	**9**
TOTAL	**100**	**100**	**100**	**100**	**100**	**100**

Q.21 × Q.17

Table 115 Downtime of robots compared with expectation by number of robots in use

column percentages of plants, mean numbers of robots per plant, approximate total numbers of robots in UK and percentages of total numbers of robots in UK

	PLANTS					ROBOTS	
	robots per plant				mean number of robots per plant	total number of robots in UK	per cent of total robots in UK
	1	2–5	6+	TOTAL			
	%	%	%	%	no.	no.	%
BASE	104	89	43	248			3208
More than expected	40	37	28	37	3	900	28
Same as expected	28	33	33	32	6	1200	38
Less than expected	24	18	30	23	5	750	24
Not answered	8	8	9	9	5	300	10
TOTAL	100	100	100	100		3208	100

Note: Figures for the total number of robot users in UK should be regarded as broadly indicative only. Particulars of the assumptions on which they are based are given in Appendix I.

Q.22

Table 116 Frequency of robot downtime by type of cause

column percentages

	Causes of downtime				
Frequency of downtime	robot itself	associated equipment*	upstream or downstream equipment	components, materials	other
BASE	248	248	248	248	248
Very frequent	5	9	3	5	1
Quite frequent	19	26	14	15	4
Occasional	53	38	34	32	4
Never	6	5	13	15	2
Not answered	17	22	36	34	89
TOTAL	100	100	100	100	100

* Attachments, guarding, feeders, storage, etc.

Q.22 × Q.3

Table 117 Frequency of robot downtime by type of cause and employment size of plant

column percentages

Frequency of cause of downtime	1–99	100–199	200–499	500–999	1000	TOTAL
BASE	31	27	68	42	78	248
Robot itself						
Very frequent	10	4	6	7	3	5
Quite frequent	10	11	19	26	23	19
Associated equipment						
Very frequent	13	0	6	12	9	9
Quite frequent	16	22	22	26	36	26
Upstream or downstream						
Very frequent	3	0	3	2	4	3
Quite frequent	13	7	15	14	17	14
Components or materials						
Very frequent	3	0	4	5	8	5
Quite frequent	16	19	18	10	14	15

Q.22 × Q.6

Table 118 Frequency of robot downtime by type of cause and industry group

column percentages

Frequency of cause of downtime	mech. eng.	elec. eng.	vehicles aircraft ships	other metal goods	plastics	other	TOTAL
BASE	57	36	55	31	23	53	248
Robot itself							
Very frequent	4	6	9	10	4	0	5
Quite frequent	23	25	22	16	22	11	19
Associated equipment							
Very frequent	7	14	9	3	4	9	9
Quite frequent	25	25	33	32	22	23	26
Upstream or downstream							
Very frequent	4	0	4	3	0	4	3
Quite frequent	5	19	13	16	26	15	14
Components or materials							
Very frequent	0	6	11	3	4	4	5
Quite frequent	14	25	13	13	17	15	15

Q.22 × Q.10

Table 119 Frequency of robot downtime by type of cause and type of feasibility study

column percentages

Frequency of cause of downtime	in-house	company in group	robot supplier	consul- tant	none	TOTAL
BASE	*93*	*18*	*75*	*35*	*24*	*248*
Robot itself						
Very frequent	2	0	8	3	13	5
Quite frequent	19	17	16	23	33	19
Associated equipment						
Very frequent	8	6	5	14	17	9
Quite frequent	30	39	19	14	38	26
Upstream or downstream						
Very frequent	3	0	0	3	8	3
Quite frequent	11	17	15	20	21	14
Components or materials						
Very frequent	4	6	3	6	17	5
Quite frequent	14	6	16	20	13	15

Q.22 × Q.17

Table 120 Frequency of downtime by type of cause and number of robots in use

*column percentages of plants, mean numbers of robots per plant,
approximate total numbers of robots in UK
and percentages of total numbers of robots in UK*

Frequency of cause of downtime	PLANTS				ROBOTS		
	robots per plant				mean number of robots per plant	total number of robots in UK	per cent of total robots in UK
	1	2–5	6+	TOTAL			
	%	%	%	%	no.	no.	%
BASE	104	89	43	248			3208
Robot itself							
Very frequent	7	7	0	5	2	100	2
Quite frequent	13	21	23	19	7	950	29
Associated equipment							
Very frequent	11	8	7	9	3	200	7
Quite frequent	22	24	44	26	8	1450	45
Upstream or downstream							
Very frequent	3	3	2	3	3	50	2
Quite frequent	14	11	21	14	6	600	19
Components or materials							
Very frequent	3	3	9	5	7	200	7
Quite frequent	14	13	21	15	5	550	17

Note: Figures for the total number of robots in UK should be regarded as broadly indicative only. Particulars of the assumptions on which they are based are given in Appendix I.

Q.23

Table 121 Benefits expected by robot users before going into production compared with those actually experienced after

column percentages

BENEFITS EXPECTED			BENEFITS EXPERIENCED		
Type of benefit	per cent of users	rank order	rank order	per cent of users	Type of benefit
BASE	248			248	BASE
Improved quality, more consistent products	75	1	1	58	Improved quality, more consistent products
Lower labour costs	69	2	2	52	Lower labour costs
Greater volume of output	61	3	3=	44	Greater volume of output
Improved work conditions, environment, safety	59	4	3=	44	Improved work conditions, environment, safety
Increased technical expertise	44	5	5	43	Increased technical expertise
Greater reliability, less down time	35	6	6	23	Better management control
Better management control	34	7	7	18	Greater reliability, less down time
Greater flexibility for product changes	27	8	8	17	Better labour relations
Lower material costs, less waste	21	9	9	16	Greater flexibility for product changes
Less capital tied up in work in progress	20	10	10	16	Lower material costs, less waste
Better labour relations	13	11	11	15	Less capital tied up in work progress
Lower equipment costs	7	12	12	4	Lower energy costs
Lower energy costs	7	12	13	3	Lower equipment costs
Other benefits	4			4	Other benefits
No benefits stated	0			19	No benefits stated
TOTAL (excl. no benefits)	476*			357*	TOTAL

* Totals add to much more than 100 because many users indicated more than one kind of benefit.

NQ.15 × NQ.8

Table 122 Benefits expected by non-users compared with benefits expected and actually experienced by existing robot users

column percentages

Type of benefit	NON-USERS				USERS	
	yes, in next two years	under considera-tion	no	TOTAL	total (expected)	total (actual)
BASE	54	187	120	363	248	248
Improved quality, more consistent products	76	64	38	56	75	58
Lower labour costs	70	63	42	57	69	52
Greater volume of output	50	46	33	42	61	44
Improved work conditions, environment, safety	37	26	16	24	59	44
Increased technical expertise	17	25	7	18	44	43
Greater reliability, less downtime	33	32	20	28	35	18
Better management control	50	33	17	30	34	23
Greater flexibility for product changes	30	25	11	21	27	16
Lower material costs, less waste	19	14	9	13	21	16
Less capital tied up in work in progress	26	19	8	17	20	15
Better labour relations	2	3	3	3	13	17
Lower equipment costs	0	2	3	2	7	3
Lower energy costs	7	5	3	5	7	4
Other benefits	4	3	2	3	4	4
No benefits stated	0	3	32	12	0	19
TOTAL (excl. no benefits)	421*	360*	212*	319*	476*	357*

* Totals add to much more than 100 because many respondents indicated more than one kind of benefit.

Q.23 × Q.3

Table 123   Benefits experienced by robot users by employment size of plant

column percentages

Benefits experienced	1–99	100–199	200–499	500–999	1000–	TOTAL
BASE	31	27	68	42	78	248
Improved quality, more consistent products	65	67	68	45	50	58
Lower labour costs	52	37	54	52	56	52
Greater volume of output	48	48	53	36	37	44
Improved work conditions, environment, safety	32	30	52	36	54	44
Increased technical expertise	36	33	52	38	46	43
Better management control	16	19	28	19	27	23
Greater reliability, less downtime	26	19	18	21	13	18
Better labour relations	13	15	24	14	17	17
Greater flexibility for product changes	10	15	21	10	18	16
Lower material costs, less waste	3	11	19	17	18	16
Less capital tied up in work in progress	13	7	10	10	24	15
Lower energy costs	0	4	9	7	1	4
Lower equipment costs	0	4	3	5	4	3
Other benefits	3	4	4	5	3	4
No benefits stated	10	22	16	24	21	19
TOTAL (excl. no benefits)	317*	313*	415*	315*	368*	357*

* Totals add to much more than 100 because many users indicated more than one kind of benefit.

Q.23 × Q.6

Table 124 Benefits experienced by robot users by industry group

column percentages

	mech. eng.	elec. eng.	vehicles aircraft ships	other metal goods	plastics	other	TOTAL
BASE	57	36	55	31	23	53	248
Improved quality, more consistent products	60	56	60	71	65	47	58
Lower labour costs	46	53	58	61	44	49	52
Greater volume of output	46	36	46	52	52	38	44
Improved work conditions, environment, safety	32	39	58	55	26	45	44
Increased technical expertise	40	36	56	52	26	36	43
Better management control	30	22	22	29	13	19	23
Greater reliability, less downtime	21	25	20	16	26	6	18
Better labour relations	7	11	26	32	17	13	17
Greater flexibility for product changes	12	11	22	23	4	13	16
Lower material costs, less waste	7	22	20	19	22	9	16
Less capital tied up in work in progress	11	19	27	10	9	9	15
Lower energy costs	5	3	4	10	4	2	4
Lower equipment costs	2	6	6	0	4	2	3
Other benefits	4	6	2	4	0	6	4
No benefits stated	16	19	11	16	22	28	19
TOTAL (excl. no benefits)	323*	345*	427*	434*	312*	294*	357*

* Totals add to much more than 100 because many users indicated more than one kind of benefit.

Q.23 × Q.13

Table 125 Benefits experienced by robot users by stage of use of first robot

column percentages

Benefits experienced	experimental, pre-production	installed for commercial production	abandoned or sold	TOTAL
BASE	*48*	*182*	*16*	*248*
Improved quality, more consistent products	25	66	56	**58**
Lower labour costs	13	63	44	**52**
Greater volume of output	17	51	44	**44**
Improved work conditions, environment, safety	23	50	50	**44**
Increased technical expertise	21	50	31	**43**
Better management control	8	27	25	**23**
Greater reliability, less downtime	4	21	19	**18**
Better labour relations	8	21	6	**17**
Greater flexibility for product changes	6	18	25	**16**
Lower material costs, less waste	6	17	31	**16**
Less capital tied up in work in progress	8	17	6	**15**
Lower energy costs	2	5	6	**4**
Lower equipment costs	0	4	0	**3**
Other benefits	4	4	0	**4**
No benefits stated	56	7	38	**19**
TOTALS (excl. no benefits)	89*	414*	343*	357*

* *Totals add to much more than 100 because many users indicated more than one kind of benefit.*

Q.23 × Q.14

Table 126 Benefits experienced by robot users by year first robot acquired

column percentages

Benefits experienced	before 1981	1981–1982	1983	1984	1985	TOTAL
BASE	33	38	41	70	57	248
Improved quality, more consistent products	85	58	59	64	37	58
Lower labour costs	73	66	51	50	33	52
Greater volume of output	70	50	44	37	32	44
Improved work conditions, environment, safety	76	61	37	46	21	44
Increased technical expertise	61	58	34	49	28	43
Better management control	42	40	15	20	12	23
Greater reliability, less downtime	27	24	17	11	16	18
Better labour relations	27	29	20	11	12	17
Greater flexibility for product changes	21	24	10	19	9	16
Lower material costs, less waste	24	29	12	19	2	16
Less capital tied up in work in progress	21	21	12	13	12	15
Lower energy costs	0	5	7	7	0	4
Lower equipment costs	3	5	5	0	5	3
Other benefits	3	3	7	3	4	4
No benefits stated	0	13	12	19	37	19
TOTAL (excl. no benefits)	533*	460*	330*	349*	223*	357*

* Totals add to much more than 100 because many users indicated more than one kind of benefit.

Q.23 × Q.4

Table 127   Benefits experienced by robot users by number of shifts a day

column percentages

Benefits experienced	NUMBER OF SHIFTS			TOTAL
	one	two	three	
BASE	*81*	*100*	*55*	*248*
Improved quality, more consistent products	57	58	64	58
Lower labour costs	36	63	58	52
Greater volume of output	37	48	47	44
Improved work conditions, environment, safety	38	50	44	44
Increased technical expertise	33	50	46	43
Better management control	15	29	24	23
Greater reliability, less downtime	14	19	22	18
Better labour relations	17	18	18	17
Greater flexibility for product changes	12	21	11	16
Lower material costs, less waste	9	20	18	16
Less capital tied up in work in progress	5	23	15	15
Lower energy costs	3	7	4	4
Lower equipment costs	0	6	4	3
Other benefits	4	4	4	4
No benefits stated	25	13	20	19
TOTAL (excl. no benefits)	280*	416*	379*	357*

*   *Totals add to much more than 100 because many users indicated more than one kind of benefit.*

Q.23 × Q.17

Table 128 Benefits experienced by robot users by number of robots in use

column percentages of plants, mean numbers of robots per plant, approximate total numbers of robots in UK and percentages of total numbers of robots in UK

Benefits experienced	PLANTS robots per plant 1	2–5	6+	TOTAL	ROBOTS mean number of robots per plant	total number of robots in UK	per cent of total robots in UK
	%	%	%	%	no.	no.	%
BASE	104	89	47	248			3208
Improved quality, more consistent products	48	61	74	58	5	2100	66
Lower labour costs	43	54	72	52	6	1950	61
Greater volume of output	37	44	63	44	5	1550	48
Improved work conditions, environment, safety	33	52	58	44	6	1650	51
Increased technical expertise	33	43	70	43	6	1750	55
Better management control	15	24	42	23	7	1100	34
Greater reliability, less downtime	14	15	30	18	6	750	24
Better labour relations	17	12	30	17	5	650	20
Greater flexibility for product changes	12	19	23	16	5	600	18
Lower material costs, less waste	6	20	30	16	7	750	23
Less capital tied up in work in progress	7	13	33	15	7	750	23
Lower energy costs	5	4	5	4	4	150	4
Lower equipment costs	3	1	9	3	8	200	7
Other benefits	5	2	5	4	4	100	3
No benefits stated	26	17	13	19	4	550	17
TOTAL (excl. no benefits)	278*	364*	544*	357*	4.3		

* Totals add to much more than 100 because many users indicated more than one kind of benefit.
Note: Figures for the total number of robots in UK should be regarded as broadly indicative only. Particulars of the assumptions on which they are based are given in Appendix I.

Q.25,26,17

Table 129 Worthwhileness and profitability of use of robots by number of robots in use

column percentages of plants, mean numbers of robots per plant, approximate total numbers of robots in UK and percentages of total numbers of robots in UK

	PLANTS				ROBOTS		
	robots per plant				mean number of robots per plant	total number of robots in UK	per cent of total robots in UK
	1	2–5	6+	TOTAL			
	%	%	%	%	no.	no.	%
BASE	104	89	43	248			3208
Use of robots has been:							
Very worthwhile	37	54	72	49	6	2000	63
Fairly worthwhile	35	31	23	32	4	850	26
Marginal or not worthwhile	20	10	0	13	2	150	4
Not answered	9	4	5	7	5	200	7
TOTAL	100	100	100	100	4.3	3208	100
Use of robots has:							
Made operations more profitable	47	66	86	61	5	2100	65
Not made operations more profitable	36	20	5	24	2	350	11
Not answered	17	13	9	15	9	750	24
TOTAL	100	100	100	100	4.3	3208	100

Note: Figures for the total number of robots in UK should be regarded as broadly indicative only. Particulars of the assumptions on which they are based are given in Appendix I.

Table 130 Factors in success: plant characteristics

row percentages

Plant characteristics	INDICATORS OF SUCCESS			INDICATORS OF LACK OF SUCCESS		
	robot very worthwhile	robot increased profit	plan to get more robots	robot marginal/not worthwhile	robot not increased profit	no plan to get more robots
TOTAL SAMPLE	49	61	61	13	24	32
Employment size						
1–99	35	48	48	**19**	**42**	**48**
100–999	53	64	59	12	26	**41**
1000+	47	60	**68**	9	15	24
Ownership of company						
UK	47	59	61	14	24	35
Overseas	**57**	**68**	62	11	23	26
Industry						
Mechanical engineering	40	54	47	17	**30**	**44**
Electrical, electronic, instrument engineering	53	56	64	**19**	**36**	25
Vehicles	**55**	**69**	**69**	5	18	25
Number of shifts						
One	36	46	47	**19**	**37**	**38**
Two	**54**	**67**	63	11	18	29
Three	**58**	**73**	**78**	7	16	18

Note: Figures in bold more than 5 percentage points above total sample.

Table 131 Factors in success: approach to introduction of robots

row percentages

Features of introduction of robots	INDICATORS OF SUCCESS			INDICATORS OF LACK OF SUCCESS		
	robot very worthwhile	robot increased profit	plan to get more robots	robot marginal/not worthwhile	robot not increased profit	no plan to get more robots
TOTAL SAMPLE	49	61	61	13	24	32
Point of decision						
Company board	48	61	59	16	24	33
Plant management	53	60	**72**	12	26	23
Feasibility study						
In-house	52	58	**67**	12	25	31
Company in group	25	**67**	**69**	6	11	19
Robot supplier	**58**	**67**	65	10	24	26
Consultant	52	49	45	16	**40**	**48**
None	46	63	38	16	21	**46**
Installation						
Turnkey by supplier	52	65	57	11	25	36
Naked robot	49	58	**66**	13	26	29
Price of first robot						
Under £20,000	52	**74**	**76**	10	14	20
Over £50,000	49	59	37	7	26	**38**
Year first robot acquired						
1980 and before	**76**	**85**	**76**	6	12	18
1983 and after	41	54	58	6	27	**37**
Workers' attitudes (after robots introduced)						
Favourable	**60**	**66**	**66**	8	16	29
Unfavourable	11	22	33	**56**	**78**	**67**

Note: Figures in bold more than 5 percentage points above total sample.

Table 132 Factors in success: number and application of robots

row percentages

	INDICATORS OF SUCCESS			INDICATORS OF LACK OF SUCCESS		
	robot very worthwhile	robot increased profit	plan to get more robots	robot marginal/not worthwhile	robot not increased profit	no plan to get more robots
TOTAL SAMPLE	49	61	61	13	24	32
Number of robots (total present installation)						
1	37	47	29	**20**	**36**	**40**
2	48	64	50	12	19	**38**
3–5	**60**	**68**	**74**	9	21	23
6 +	**72**	**86**	**77**	0	5	19
Application area (plants using robots on application specified)						
Arc welding	47	58	51	12	28	**40**
Spot welding	**62**	**69**	**62**	0	8	15
Painting, coating	**58**	**78**	**72**	11	11	19
Glueing, sealing	**59**	59	**71**	0	18	24
Assembly	46	50	**69**	12	24	26
Handling	53	**71**	**74**	16	15	26
Grinding	50	**68**	**79**	7	7	21
Injection moulding	**55**	**70**	**85**	10	20	10
Press loading	39	**77**	**77**	0	8	15
Machine loading	51	**78**	**62**	11	11	29

Note: Figures in bold more than 5 percentage points above total sample.

Q.00

Table 133 Factors in success: difficulties and benefits experienced

row percentages

	INDICATORS OF SUCCESS			INDICATORS OF LACK OF SUCCESS		
	robot very worthwhile	robot increased profit	plan to get more robots	robot marginal/not worthwhile	robot not increased profit	no plan to get more robots
TOTAL SAMPLE	49	61	61	13	24	32
Difficulties experienced						
High development costs	38	56	48	**19**	**41**	**46**
Inadequate after-sales support	40	63	56	**20**	28	**37**
Installation problems	46	59	58	15	**32**	35
Lack of specialist expertise	41	62	62	**23**	**29**	33
Insufficient reliability	34	56	56	**25**	**34**	**41**
High equipment costs	42	63	58	11	28	**39**
More downtime than expected	34	45	51	**25**	**45**	**46**
No difficulty reported	**62**	60	63	6	12	26
Benefits experienced						
Improved quality, consistency	**58**	**74**	**66**	8	19	29
Lower labour costs	**56**	**78**	63	8	15	33
Greater volume of output	**63**	**85**	**68**	5	10	27
Improved work conditions, safety	**57**	**77**	**65**	5	16	**36**
Increased technical expertise	**57**	**79**	**64**	7	16	33
Better management control	**69**	**83**	**69**	2	12	29
Greater reliability, less downtime	**71**	**84**	**76**	2	9	20
No benefits reported	26	22	48	**24**	**33**	33

Note: Figures in bold more than 5 percentage points above total sample.

Q.33 × Q.6

Table 134 Plans to acquire more robots in next two years by industry group

column percentages

Plans in next two years	mech. eng.	elec. eng.	vehicles aircraft ships	other metal goods	plastics	other	TOTAL
BASE	*81*	*52*	*73*	*35*	*30*	*65*	*326*
Whether plan to acquire more robots:							
Yes	49	64	67	63	80	59	60
No	41	27	27	31	20	37	33
Not answered	10	10	6	6	0	5	6
TOTAL	100	100	100	100	100	100	100
Number more:							
1	9	15	10	26	17	19	14
2	19	14	25	11	13	11	16
3–5	10	19	11	9	13	8	11
6–10	3	4	7	3	17	5	6
over 10	1	2	8	3	3	3	4
Mean number more robots:							
per plant planning more	3	3	6	3	4	4	4.4
all existing users	2	2	4	2	4	3	2.6
Total number more robots planned in UK	300	250	700	150	250	400	1950

Note: Figures for the total number of robots and robot users in UK should be regarded as broadly indicative only. Particulars of the assumptions on which they are based are given in Appendix I.

Q.33 × Q.1

Table 135 Plans to acquire more robots in next two years by type of company

column percentages

Plans in next two years	UK company	overseas company	TOTAL	Total no. users in UK
BASE	197	47	248	
Whether plan to acquire more robots:				
Yes	60	62	60	440
No	35	26	33	240
Not answered	6	13	6	
TOTAL	100	100	100	740
Number more:				
1	17	11	14	100
2	15	13	16	120
3–5	9	19	11	80
6–10	6	6	6	44
over 10	3	4	4	30
over 20	0	2		
Mean number more robots:				
per plant planning more	4	6	4·4	
all existing users	2	4	2·6	
Total number more robots in UK	*1400*	*550*	*1950*	

Note: Figures for the total number of robots and robot users in UK should be regarded as broadly indicative only. Particulars of the assumptions on which they are based are given in Appendix I.

Q.33 × Q.3

Table 136 Plans to acquire more robots in next two years by employment size of plant

column percentages

Plans in next two years	1–99	100–199	200–499	500–999	1000–	TOTAL
BASE	*40*	*39*	*91*	*50*	*104*	*326*
Whether plan to acquire more robots:						
Yes	50	51	62	58	66	60
No	48	44	31	34	27	33
Not answered	3	5	8	8	7	6
TOTAL	100	100	100	100	100	100
Number more:						
1	18	18	17	18	7	14
2	18	10	20	14	14	16
3–5	5	15	10	10	14	11
6–10	0	5	6	8	7	6
over 10	0	0	1	2	10	4
Mean number more robots:						
per plant planning more	2	3	3	3	8	4·4
all existing users	1	2	2	2	5	2·6
Total number more robots in UK	*100*	*150*	*350*	*200*	*1150*	*1950*

Note: Figures for the total number of robots in UK should be regarded as broadly indicative only. Particulars of the assumptions on which they are based are given in Appendix I.

Q.33 × Q.4

Table 137 Plans to acquire more robots in next two years by number of shifts a day

column percentages

Plans in next two years	one shift	two shifts	three shifts	TOTAL
BASE	*81*	*100*	*55*	*248*
Whether plan to acquire more robots:				
Yes	47	63	78	60
No	44	29	18	33
Not answered	9	8	4	6
TOTAL	100	100	100	100
Number more:				
1	20	13	16	14
2	11	22	11	16
3–5	3	10	24	11
6–10	3	5	13	6
over 10	0	6	4	4
Mean number more robots:				
per plant planning more	2	5	4	4.4
all existing users	1	3	3	2.6
Total number more robots in UK	*250*	*1000*	*550*	*1950*

Note: Figures for the total number of robots and robot users in UK should be regarded as broadly indicative only. Particulars of the assumptions on which they are based are given in Appendix I.

Q.33 × Q.14

Table 138 Plans to acquire more robots in next two years by year first robot acquired

column percentages

Plans in next two years	before 1981	1981–1982	1983	1984	1985	TOTAL
BASE	*33*	*38*	*41*	*70*	*57*	*248*
Whether plan to acquire more robots:						
Yes	76	63	51	57	63	60
No	18	26	46	39	28	33
Not answered	6	11	2	4	9	6
TOTAL	100	100	100	100	100	100
Number more:						
1	15	11	7	17	26	14
2	21	18	5	13	19	16
3–5	12	8	20	10	5	11
6–10	6	11	12	1	2	6
over 10	16	5	2	4	0	4
Mean number more robots:						
per plant planning more	6	5	5	4	2	4·4
all existing users	5	3	2	2	1	2·6
Total number more robots in UK	500	400	300	500	200	1950

Note: Figures for the total number of robots and robot users in UK should be regarded as broadly indicative only. Particulars of the assumptions on which they are based are given in Appendix I.

Q.33 × Q.17

Table 139 Plans to acquire more robots by number of robots in use now

column percentages of plants, mean numbers of robots per plant, approximate total numbers of robots in UK and percentages of total numbers of robots in UK

	PLANTS					ROBOTS	
	robots per plant				mean number of robots per plant	total number of robots in UK now	per cent of total robots in UK now
	1	2–5	6+	TOTAL			
	%	%	%	%	no.	no.	%
BASE	140	121	52	326			3208
Whether plans to acquire more robots:							
Yes	50	64	79	60	5	2150	67
No	44	30	17	33	3	950	29
Not answered	6	7	4	1	3	100	4
TOTAL	100	100	100	100	4.3	3208	100
Number more:							
1	21	11	0	14	1	150	5
2	14	21	13	16	3	400	12
3–5	6	13	15	11	5	400	13
6–10	1	5	21	6	7	350	11
Over 10	0	3	13	4	20	400	13
Mean number more:							
Per plant planning more	2	4	9	4.4			
All existing users	1	5	8	2.6			
Total number more in UK	300	750	850	1950			

Note: Figures for the total number of robots and robot users in UK should be regarded as broadly indicative only. Particulars of the assumptions on which they are based are given in Appendix I.

Q.17 × Q.33

Table 140 Number of robots per plant now by number of additional robots planned in next two years

column percentages of plants, mean numbers of robots per plant, approximate total numbers of robots in UK and percentages of total numbers of robots in UK

	PLANTS				ROBOTS		
	additional robots per plant				mean number more robots per plant	total additional robots in UK	per cent of additional robots in UK
Number of robots per plant now	1	2–5	6+	TOTAL			
	%	%	%	%	no.	no.	%
BASE	45	89	30	326			1950
1	67	33	3	43	2	300	16
2	20	16	10	18	3	250	12
3–5	9	30	23	19	5	500	28
6–10	0	11	30	8	6	250	15
Over 10	0	6	30	8	13	550	29
Not answered	4	4	3	4			
TOTAL	100	100	100	100	4.4	1950	100
Mean number per plant	4	4	12	4.3			

Note: Figures for the total number of robots and robot users in UK should be regarded as broadly indicative only. Particulars of the assumptions on which they are based are given in Appendix I.

Q.33 × Q.21

Table 141 Plans to acquire more robots in next two years by whether robot downtime more or less than expected

Plans in next two years	DOWNTIME			TOTAL
	more than expected	about same	less than expected	
BASE	*91*	*79*	*56*	*326*
Whether plans to acquire more robots:				
Yes	51	67	71	60
No	46	27	23	33
Not answered	3	6	5	6
TOTAL	100	100	100	100
Number more:				
1	12	20	14	14
2	15	16	20	16
3–5	9	13	9	11
6–10	4	6	9	6
over 10	4	5	2	4
Mean number more:				
per plant planning more	*5*	*4*	*3*	*4.4*
all existing users	*2*	*3*	*2*	*2.6*
Total number more in UK	*700*	*750*	*400*	*1950*

Q.33 × Q.25,26

Table 142 Plans to acquire more robots in next two years by extent use of robots worthwhile and profitable

column percentages

Plans in next two years	very worthwhile	fairly worthwhile	marginal or not worthwhile	profitable	not profitable	TOTAL
BASE	122	78	31	152	60	326
Whether plans to acquire more robots:						
Yes	79	50	29	74	37	60
No	16	46	65	21	60	33
Not answered	5	4	7	5	3	6
TOTAL	100	100	100	100	100	100
Number more:						
1	19	13	10	18	8	14
2	20	13	10	19	10	16
3–5	15	8	0	15	3	11
6–10	8	4	3	6	7	6
Over 10	2	8	0	4	2	4
Mean number more:						
Per plant planning more	4	5	2	4	4	4.4
All existing users	3	3	1	3	1	2.6
Total number more in UK	1150	650	50	1450	250	1950

NQ.9 × Q.3

Table 143 Non-users' robot feasibility studies by employment size of plant

column percentages

Feasibility study	1–99	100–199	200–499	500–999	1000–	TOTAL
BASE	*114*	*57*	*88*	*62*	*41*	*363*
No study undertaken or planned	79	65	60	47	42	63
Study planned	8	9	11	11	20	11
Study in progress	0	5	5	8	7	4
Study completed	11	19	23	32	32	21
Did not recommend use of robots	9	14	9	16	20	12
Recommended robots	4	7	10	18	10	9
TOTAL	100	100	100	100	100	100

NQ.9 × Q.4

Table 144 Non-users' robot feasibility studies by number of shifts a day

column percentages

Feasibility study	one	two	three	TOTAL
BASE	*153*	*100*	*45*	*363*
No study undertaken or planned	67	52	53	63
Study planned	14	11	9	11
Study in progress	3	7	2	4
Study completed	15	29	31	21
Did not recommend use of robots	7	15	16	12
Recommended robots	6	15	13	9
TOTAL	100	100	100	100

NQ.9 × Q.6

Table 145 Non-users' robot feasibility studies by industry group

column percentages

Feasibility study	mech. eng.	elec. eng.	vehicles aircraft ships	other metal goods	plastics	other	TOTAL
BASE	*104*	*83*	*34*	*37*	*22*	*104*	*363*
No study undertaken or planned	66	60	47	46	77	68	63
Study planned	7	18	24	14	5	7	11
Study in progress	7	2	3	8	5	3	4
Study completed	20	18	27	30	9	20	21
Did not recommend use of robots	11	11	9	19	9	12	12
Recommended robots	7	6	15	14	5	10	9
TOTAL	100	100	100	100	100	100	100

NQ.8 × Q.9

Table 146 Non-users' plans to use robots in next two years by whether feasibility study undertaken

column percentages

	FEASIBILITY STUDY						
Plans to use robots	None planned	planned	in progress	completed	did not recommend robots	recommended robots	TOTAL
BASE	*227*	*39*	*15*	*77*	*44*	*32*	*461*
Yes, in next two years	4	49	27	27	7	50	15
Under consideration	49	51	73	56	59	50	49
No	47	0	0	17	32	0	35
Not answered	0	0	0	0	0	0	1
TOTAL	100	100	100	100	100	100	100

Q.8 × Q.3

Table 147 Non-users' plans to use robots in next two years by employment size of plant

column percentages

Plans to use robots	1–99	100–199	200–499	500–999	1000–	TOTAL
BASE	*151*	*70*	*115*	*74*	*50*	*461*
Whether plans to use robots:						
Yes, in next two years	8	16	17	22	20	15
Under consideration	37	47	54	55	66	49
No	53	36	29	23	12	35
Not answered	2	1	1	0	2	1
TOTAL	100	100	100	100	100	100
Number planned in next two years:						
1	1	6	5	8	2	4
2	3	9	5	5	6	5
Over 2	2	3	3	3	6	4
Mean number more:						
Per plant planning more	*4*	*2*	*3*	*3*	*3*	*2.9*
All potential users	*0*	*0*	*0*	*1*	*1*	*0.4*

NQ.8 x Q.2

Table 148   Non-users' plans to use robots in next two years by number of plants of company

column percentages

Plans to use robots	1	2	3–5	5–	TOTAL
BASE	*149*	*73*	*58*	*80*	*461*
Whether plans to use robots:					
Yes, in next two years	11	15	12	24	15
Under consideration	42	58	64	58	49
No	46	27	24	19	35
Not answered	1	0	0	0	1
TOTAL	100	100	100	100	100
Number planned in next two years:					
1	3	1	2	6	4
2	4	6	7	6	5
Over 2	3	3	0	8	4
Mean number more:					
Per plant planning more	3	3	2	3	2.9
All potential users	0	1	0	1	0.4

NQ.8 × Q.4

Table 149 Non-users' plans to use robots in next two years by number of shifts a day

column percentages

Plans to use robots	one	two	three	TOTAL
BASE	153	100	45	461
Whether plans to use robots:				
Yes, in next two years	13	23	20	15
Under consideration	51	58	58	49
No	36	19	20	35
Not answered	0	0	2	1
TOTAL	100	100	100	100
Number planned in next two years:				
1	3	5	4	4
2	6	7	4	5
over 2	2	8	4	4
Mean number more:				
per plant planning more	2	3	4	2.9
all potential users	0	1	1	0.4

NQ.8 × Q.6

Table 150 Non-users' plans to use robots in the future by industry group

column percentages

Plans to use robots	mech. eng.	elec. eng.	vehicles aircraft ships	other metal goods	plastics	other	TOTAL
BASE	142	103	42	46	30	131	461
Whether plans to use robots:							
Yes, in next two years	13	17	21	24	17	9	15
Under consideration	46	54	52	48	50	50	49
No	40	28	21	28	30	40	35
Not answered	1	1	5	0	3	1	1
TOTAL	100	100	100	100	100	100	100
Number planned in next two years:							
1	4	3	5	9	7	2	4
2	5	5	10	9	0	3	5
over 2	2	3	7	2	10	3	4
Mean number more:							
per plant planning more	3	3	2	3	7	3	2.9
all potential users	0	0	1	1	1	0	0.4

Table 151 Growth in total number of robots and robot users in UK on alternative assumptions by industry group

numbers of robots and robot users

		mech. eng.	elec. eng.	vehicles	other metal goods	plastics	other	TOTAL
ROBOT USERS								
Existing users		180	110	160	80	70	140	740
Non-users expecting to use robots in next two years								
	Low	40	40	20	20	10	30	160
	High	200	190	100	120	60	130	800
Total users in two years								
	Low	220	150	180	100	80	170	900
	High	380	300	260	200	130	270	1540
ROBOTS								
Existing robots		450	300	1350	200	450	450	3200
Existing users' expectations of additional robots in next two years								
	Low	100	100	200	50	50	150	650
	High	300	250	700	150	200	350	1950
Non users' expectations of robots in next two years								
	Low	100	100	50	50	50	50	400
	High	450	500	250	300	250	350	2100
Total robots in two years								
	Low	650	500	1600	300	550	650	4250
	High	1200	1050	2300	650	900	1150	7250

Note: The above figures are indicative only. The assumptions on which they are based are explained in Appendix I and their implications are discussed in the text of the report.

Table 152 Growth in total number of robots and robot users in UK on alternative assumptions by employment size of plant

numbers of robots and robot users

		1–99	100–199	200–499	500–999	1000–	TOTAL
ROBOT USERS							
Existing users		90	90	210	110	240	740
Non-users expecting to use robots in next two years							
	Low	30	30	40	40	20	160
	High	140	130	220	190	120	800
Total users in two years							
	Low	120	120	250	150	260	900
	High	230	220	430	300	360	1540
ROBOTS							
Existing robot		150	150	800	400	1700	3200
Existing users' expectations of additional robots in next two years							
	Low	50	50	100	50	400	650
	High	100	150	350	200	1150	1950
Non-users' expectations of robots in next two years							
	Low	100	50	100	100	50	400
	High	450	250	550	500	350	2100
Total robots in two years							
	Low	300	250	1000	550	2150	4250
	High	700	550	1700	1100	3200	7250

Note: The above figures are indicative only. The assumptions on which they are based are explained in Appendix I and their implications are discussed in the text of the report.

Q.33 × Q.34

Table 153 Plans to acquire more robots in next two years by degree of sophistication of new robot

column percentages

Plans in next two years	more sophisticated	about same	less sophisticated	TOTAL
BASE	73	114	19	326
Whether plans to acquire more robots:				
Yes	73	69	74	60
No	23	27	16	33
Not answered	4	4	11	6
TOTAL	100	100	100	100
Number more:				
1	18	18	26	14
2	21	14	26	16
3–5	12	13	5	11
6–10	7	8	0	6
over 10	4	4	5	4
Mean number more:				
per plant planning more	4	4	3	4.4
all existing users	3	3	2	2.6
Total number more in UK	700	1050	100	1950

Q.35 × Q.3

Table 154 Improvements robot users consider would help effective use in the future by employment size of plant

column percentages

	1–99	100–199	200–499	500–999	1000–	TOTAL
BASE	31	27	68	42	78	248
Cheaper robots	71	67	78	67	67	70
Easier programming	48	52	57	52	63	56
Cheaper associated equipment	36	67	50	45	63	53
More government support	68	63	57	48	35	50
Better sensors	52	44	46	43	58	50
Less need for special skills	36	52	41	48	41	43
An upturn in the economy	42	44	47	36	36	41
Easier maintenance	26	26	38	43	45	39
Greater speed	29	44	49	33	28	37
More intelligence	26	30	32	29	45	35
Better reliability	32	26	35	41	36	35
More accuracy	19	37	29	24	37	31
Greater versatility	29	26	32	24	32	29
More UK based manufacturers	19	33	34	24	30	29
Better after-sales service	10	19	31	31	27	26
Easier finance for investment	19	26	31	17	15	21
Lower operating costs	18	33	25	17	23	21
Heavier pay load	10	15	15	7	35	19
Easier installation	0	15	19	33	19	19
Quicker delivery	23	11	12	5	8	11
Smaller lighter robots	7	11	10	12	12	11
Other	0	4	2	10	6	4
Not answered	0	0	3	5	1	2

Q.35 × Q.6

Table 155 Improvements robot users consider would help effective use in the future by industry group

column percentages

	mech. eng.	elec. eng.	vehicles aircraft ships	other metal goods	plastics	other	TOTAL
BASE	*57*	*36*	*55*	*31*	*23*	*53*	*248*
Cheaper robots	68	58	80	84	65	68	70
Easier programming	58	61	56	42	61	55	56
Cheaper associated equipment	54	50	58	48	44	57	53
More government support	63	58	47	58	44	36	50
Better sensors	46	69	66	39	35	42	50
Less need for special skills	40	50	44	45	44	45	43
An upturn in the economy	51	33	42	45	39	25	41
Easier maintenance	35	33	44	42	26	36	39
Greater speed	37	33	20	45	44	47	37
More intelligence	35	39	53	19	26	28	35
Better reliability	35	44	40	36	30	23	35
More accuracy	32	39	36	23	35	25	31
Greater versatility	26	36	35	36	39	25	29
More UK based manufacturers	30	31	38	32	13	21	29
Better after-sales service	25	33	26	36	26	15	26
Easier finance for investment	25	17	27	23	26	17	21
Lower operating costs	26	17	26	26	13	11	21
Heavier payload	12	11	33	19	9	23	19
Easier installation	12	22	18	13	30	21	19
Quicker delivery	12	14	9	10	17	4	11
Smaller, lighter robots	2	11	9	7	13	21	11
Other	4	8	2	0	0	0	4
Not answered	0	3	0	0	4	4	2

Table A Basis of postal survey: sources of sample and rates of response

Figures in italics include additional response from abbreviated questionnaires

Source of sample*	total questionnaires despatched**	completed questionnaires received			response rate
		existing robot users	potential robot users	total	
	no.	no.	no.	no.	%
KNOWN ROBOT USERS					
DTI: firms with support grants for use of robots	171	109 *135*	0 *0*	109 *135*	64 *79*
PSI: plants using robots in 1985 survey of microelectronics in industry	65	36 *43*	1 *5*	37 *48*	57 *74*
Robot suppliers: plants to which robots supplied	62	24 *30*	0 *4*	24 *34*	39 *55*
Total	298	169 *208*	1 *9*	170 *217*	57 *73*
POTENTIAL ROBOT USERS					
Automan: visitors to 1985 exhibition	952	58 *87*	317 *397*	375 *484*	39 *51*
PSI: plants expecting to use robots by 1987 in 1985 survey of microelectronics in industry	59	3 *8*	24 *31*	27 *39*	46 *66*
BRA: company members believed to be potential users	74	18 *23*	21 *24*	39 *47*	53 *64*
Total	1085	79 *118*	362 *452*	441 *570*	41 *53*
TOTAL	1383	248 *326*	363 *461*	611 *787*	44 *57*

* There were considerable duplications between the lists. The figures in each category given above exclude those already included in the categories above it.

** figures for despatches exclude a total of 114 questionnaires sent to addresses which turned out to be incorrect or to firms which were not possible robot users or were unsuitable for inclusion for other reasons.

Table B: Sample of robot users by plant size and industry

Number of plants

Industry	1-19	20-49	50-99	100–199	200–499	500–999	1000–4999	5000–	NA	TOTAL
Food, drink, tobacco	0	0	0	0	1	1	4	1	0	7
Chemicals, metals	0	1	0	2	3	1	5	0	0	12
Mechanical engineering	1	7	6	10	19	7	7	0	0	57
Electrical, electronic, instrument engineering	1	0	2	2	8	14	8	1	0	36
Vehicles, aircraft, shipbuilding	0	0	3	2	12	4	18	16	0	55
Other metal goods	1	2	3	3	11	4	7	0	0	31
Textiles	0	0	0	0	0	0	0	1	0	1
Clothing, footwear, leather, fur	0	0	0	1	1	0	0	1	0	3
Paper, printing, publishing	0	0	0	2	0	1	2	0	0	5
Bricks, pottery, glass, cement, wood products	1	0	1	2	0	5	2	0	0	11
Plastic products	0	1	3	3	13	2	0	1	0	23
Other manufacturing	0	0	0	2	1	2	3	1	1	10
Non-manufacturing	1	0	0	0	0	2	1	1	0	5
Not answered	0	0	0	2	1	0	0	0	1	4
TOTAL	5	11	15	27	68	42	56	22	2	248

Table C Sample of non-users by plant size and industry

numbers of plants

Industry	1-19	20-49	50-99	100-199	200-499	500-999	1000-4999	5000-	NA	TOTAL
Food, drink, tobacco	0	2	1	0	5	4	2	2	0	16
Chemicals, metals	1	2	4	4	6	5	2	1	0	25
Mechanical engineering	16	15	9	22	19	15	7	1	0	104
Electrical, electronic, instrument engineering	13	7	6	11	18	17	10	0	1	83
Vehicles, aircraft, shipbuilding	1	2	5	6	5	5	8	2	0	34
Other metal goods	1	2	2	5	15	10	2	0	0	37
Textiles	0	0	2	1	3	0	1	0	0	7
Clothing, footwear, leather, fur	0	0	0	1	1	0	0	0	0	2
Paper, printing, publishing	5	1	0	1	5	0	1	0	0	13
Bricks, pottery, glass, cement, wood products	3	2	3	6	7	6	2	0	0	29
Plastic products	2	2	5	3	7	2	1	0	0	22
Other manufacturing	0	0	0	1	1	0	0	0	0	2
Non-manufacturing	7	1	0	1	0	1	0	1	0	11
Not answered	1	0	0	0	0	0	0	0	0	1
TOTAL	47	34	33	57	88	62	35	6	1	363

Table D Comparison of the plant size distribution of the robot users in the survey sample with the UK distribution of robot users in the PSI survey of microelectronics in industry.

percentage of sample in each employment size

Employment size	PSI industry survey weighted for all UK	Total survey sample	survey sample sources				
			DTI support scheme	PSI industry survey	robot suppliers	Automan exhibition	BRA
SAMPLE BASE	80	248* (326*)	109 (135)	53 (66)	56 (66)	138 (184)	18 (23)
1–199	22.3	23.6 (24.4)	29	8	5	20	17
200–499	27.9	27.6 (28.1)	32	11	30	29	28
500–999	17.2	17.1 (15.4)	12	23	23	17	28
1000–	32.5	31.7 (32.1)	26	59	41	33	28
TOTAL	100	100	100	100	100	100	100

* The total sample used is less than the sum of the samples from different sources because of the elimination of duplications where the same user occurs on the lists from more than one source.
Figures in brackets include the supplementary mini-questionnaire.

Table E Comparison of the distribution by industry of the robot users in the survey sample with the UK distribution of robot users in the PSI survey of microelectronics in industry

percentage of sample in each industry group

Industry group	PSI industry survey weighted for all UK	Total survey sample	survey sample sources				
			DTI support scheme	PSI industry survey	robot suppliers	Automan exhibition	BRA
SAMPLE BASE	80	248*	109	53	56	138	18
Food, drink, tobacco	2	3	4	4	0	3	0
Chemicals, metals	6	5	5	2	4	7	11
Mechanical engineering	20	23	27	11	18	22	28
Electrical, electronic, instrument engineering	22	15	16	23	16	17	11
Vehicles, aircraft, shipbuilding	24	22	16	38	27	21	22
Other metal goods	2	13	16	11	16	9	6
Textiles	2	0	1	0	2	1	0
Clothing, footwear, leather, fur	0	1	1	2	2	1	0
Paper, printing, publishing	2	2	2	2	2	2	0
Other	20	20	18	9	15	20	28
TOTAL	100	104**	106**	102**	102**	103**	106**

* The total sample used is less than the sum of the samples from different sources because of the elimination of duplications where the same user occurs on the lists from more than one source.
** Totals add to slightly more than 100% because a few respondents gave more than one industry for their company's main activity.